COMMON GROUND

NEW LITERACIES, NEW LEARNING

CYBER-SOCIAL PEDAGOGIES IN THE BRAZILIAN CONTEXT

RODRIGO ABRANTES DA SILVA

NEW LITERACIES, NEW LEARNING

CYBER-SOCIAL PEDAGOGIES IN THE BRAZILIAN CONTEXT

NEW LITERACIES, NEW LEARNING

CYBER-SOCIAL PEDAGOGIES IN THE BRAZILIAN CONTEXT

RODRIGO ABRANTES DA SILVA

SÃO PAULO

2025

COMMON GROUND

First published in 2025
as part of the **Information, Medium & Society Book Imprint**

Common Ground Research Networks
University of Illinois Research Park
2001 South First St, Suite 201 L
Champaign, IL 61820 USA

Library of Congress Cataloging-in-Publication Data Forthcoming

ISBN:978-1-966214-62-5 (HBK)

ISBN:978-1-966214-63-2 (PBK)

ISBN: 978-1-966214-64-9 (PDF)

DOI: 10.18848/978-1-966214-64-9/CGP

DEDICATION

For Carol and Sebastião.

ACKNOWLEDGMENTS

First and foremost, I express my deepest gratitude to my family, whose unwavering support and constant encouragement were fundamental throughout the entire process of the research that led to this book. Without you, this journey would not have been possible.

To my adviser, Walkyria Monte Mor, I extend my most sincere thanks for the trust you placed in me, for your contagious enthusiasm, and for the inspiring collaboration. Your guidance was essential to the completion of this work.

I am immensely grateful to Professors Bill Cope and Mary Kalantzis for welcoming me into their research group and for all the opportunities for learning and growth they provided.

To my new friends, Vânia Castro and Jailine Maiara Farias, I am grateful for the enriching experiments with CGScholar, which greatly contributed to the results of this research.

To Professor Suzanna Mizan, I extend my thanks for the collaborative research we conducted, always valuing dialogue and affection in education.

To my friends from the National Literacies Project and the Cyber-Social Learning Group, my heartfelt thanks for the support and encouragement throughout the process, especially to those who read my work and provided feedback.

To my friends at the Innovation and Technology Center of Dante Alighieri School, who were important companions during the preparation of this book, my gratitude for the constant collaboration and support.

Lastly, to Rosali Figueiredo, my gratitude for the friendship and for helping me adapt my doctoral thesis into this book.

CONTENTS

ABOUT THE BOOK

In this book, I present a journey of experimentation, inquiries, and discoveries regarding my relationship with education in a digital environment. I began this path as a history teacher in primary and secondary schools, then became a tutor for my colleagues at the institution, got involved with researchers and digital inclusion programs supported by non-governmental organizations, until I decided to study the subject within the scope of academic research. This journey inspired my doctoral research, which I presented this year at the University of São Paulo.

The book provides an overview of the main findings from the doctoral thesis, among them the conviction that the digital ecosystem offers functionalities that can foster critical, diverse, and inclusive education, provided it is presented in the form of a cyber-social design, focused on multiliteracies and new pedagogical practices, such as peer learning (feedback) and process-oriented assessment (not just focused on the outcome). In this ecosystem, the relationship between teacher and student is horizontal, and new knowledge is built throughout the learning process, enabling students to expand their knowledge repertoire, recognize their own context, adopt a critical stance in relation to society, and become agents in building their life projects.

Of course, this requires a willingness from society, schools, teachers, and the state to move in this direction. However, it is important for everyone to know that, from a pedagogical perspective, experiences with education in a digital environment can be transformative, despite the current predominance of the exacerbation of conventional education's massification through market platforms.

To validate this idea, we monitored teaching and learning interactions on an educational platform developed in the United States by the College of Education at the University of Illinois Urbana-Champaign and implemented in Brazil for undergraduate and graduate students through a partnership with the Department of Modern Languages at FFLCH-USP (Faculty of Philosophy, Languages and Human Sciences at the University of São Paulo). This platform, CGScholar, features a cyber-social design.

This concept, along with other central perspectives adopted in this work—including literacies and multiliteracies—is presented in Chapters 1 and 2. In the first chapter, I provide an overview of the development of digital technology within the context of education, introducing these concepts minimally and exploring the origins of the idea of feedback, grounded in cybernetics. In Chapter 2, I describe the transition from conventional educational environments to a cyber-social dynamic, where the human component (and the related subjectivities) can harness the technological structure and excel in its interaction with the machine. In this sense, I reflect on the development of the Web and emphasize that the evolution of its language, as well as the ability to operate simultaneously with different forms of expression (texts, videos, and photos, among others) from anywhere in the world in easily convertible languages, has expanded pedagogical possibilities.

In other words, the Web and its entire repertoire of functionalities, information exchange, and knowledge sharing open what researchers in the field of Cyber-Social education define as new windows of learning opportunities (*affordances*), namely the ubiquitous relationship between the experience of education and digital apparatuses; active creation of knowledge; multimodal representations of meaning; recursive feedback, which feeds back into new flows of information exchange and impressions between teacher and student, and between students; metacognition (self-reflection accompanies the acquisition and production of knowledge); and differentiated and social learning, where the diversity of students becomes the raw material of the process.

How did we get here? In Chapter 3, I outline a brief methodological journey of my research, which is important in the context of this book as it demonstrates how previous experiences in the classroom and tutoring helped me raise questions and test the windows of opportunity provided by the CGScholar platform. It is also essential to note that my study, as well as those of numerous colleagues who participated in the partnership between the University of Illinois Urbana-Champaign College of Education and USP, could only be produced because we had access to a platform that we would not have been able to develop under the current conditions of academic research in Brazil.

Access to CGScholar allowed us to observe the feasibility of creating infrastructures for digital ecosystems that foster more inclusive and critical education. The description of the platform's functionalities, as well as the dynamics and learning resources it provides, is also covered in Chapter 3.

In Chapters 4 and 5 I present case studies, examples extracted from communities of undergraduate and graduate courses in Brazil, which may inspire solutions for Basic Education. I also offer a brief discussion on innovative pedagogical practices, such as peer review and the importance of recurring feedback, not only from teacher-tutors to students but also among the students themselves. Finally, I present other assessment possibilities for students, both individually and in groups, including the significant contribution of Artificial Intelligence tools.

It is true that, in a society as socially and economically unequal as Brazil, the implementation of cyber-social ecosystems in education faces a very challenging scenario, as discussed in Chapter 6. However, in this book, I aim to demonstrate that there is sufficient digital infrastructure and theoretical and empirical resources to invest in this path. We can reimagine our relationship with technology, addressing the expectations and needs of new generations of students, engaging digital natives in meaningful learning with implications for their real-world experiences.

Rodrigo Abrantes da Silva

FOREWORD

When revisiting the multiliteracies framework, created and discussed by a group of educators and researchers in 1996, two of them, Bill Cope et al., emphasize the project's alignment with a necessary educational justice program, providing "social access for historically marginalized individuals and communities," serving as "a point of passage for a broader social transformation, based on a deeper ethical and systemic agenda of equality" (2023, p. 3). Regarding the challenge of integrating the literate society with the digital society, I agree with their proposed goals and inquiries:

> How can we support, through education, the formation of social agents able to address, individually and collectively, with science and humanity, the manifold crises of our times? How can learners make sense of the world then assume their responsibility to act in it? (Cope et al., 2023, p. 3)

New Literacies, New Learning: Cyber-Social Pedagogies in the Brazilian Context addresses these concerns. This is undoubtedly an innovative work by Rodrigo Abrantes da Silva. It is the result of his doctoral research, which I had the great pleasure of supervising and from which I also learned a great deal. The CGScholar platform, created by Bill Cope, Mary Kalantzis, and collaborators from their research group in Illinois, is a very relevant part of Silva's research.

As the author himself explains, by researching a literacy practice through a platform not linked to the Big Five (Google, Apple, Meta, Amazon, and Microsoft, also known as GAFAM), he came to understand and investigate the operationalization and educational potential of the Scholar platform. Lacking a commercial character, Scholar receives support from research agencies and the U.S. government. It offers a reimagined perspective on education, focusing on studies, research, and exchanges. Its design envisions education as a social process of knowledge construction in which teachers are crucial for designing the courses.

On the platform, teachers take on roles such as fostering trust through interactions and collaborations among students; analyzing data from peer review processes; and reviewing and publishing the work produced. The social practice of teaching and learning is altered according to the digital possibilities provided by the platform.

In this way, on Scholar, the digital has an amplifying role in education by proposing—and making it possible—that it go beyond its technical function. That is, the specific characteristics of digital knowledge facilitate an ontoepistemological expansion in learning. In the conservative view of language still observed in many schools and universities, the predominant epistemology involves, for example, understanding parts to form a whole (with didactic motivation, the whole is fragmented for this purpose), linearity and sequence, gradation, and hierarchy of knowledge. In contrast, the digital enables expansions by creating space for other forms of peer interaction, understanding the multimodality of language, and recognizing alternative processes of knowledge construction.

This "opening up" of digital language/knowledge acts as a "break from the mold" or a "cutter of corset strings," according to an analogy we often use (Monte Mor et al., 2024), based on a description by Gee and Hayes (2011):

> Contemporary society, shaped by the digital age, interacts with diverse logics and epistemologies that go beyond traditional writing technologies such as computers, notebooks, and apps, extending into broader ontoepistemological realms. Rather than replacing analogue writing technologies, the advent of digital mediums has expanded various languages—illustrated by multimodalities—by loosening the "corsets of normativity" [referencing Gee and Hayes (2011), who use the analogy of language to human breath: conventional, normatized language is seen as constricting human breath within corsets, imposing rules and creating a model of homogeneity. The digital seems to have untied these tight knots, allowing breath to also embrace its pluralities]. (Monte Mor et al., 2024, p. 227)

Recent studies show that as young people engage with the digital world, they develop and incorporate new ways of learning—many of which had previously existed informally outside the formal educational scope—integrating them with established social and academic methods. Additionally, they come to understand that language extends beyond written and normative processes. They recognize language socially, acknowledging its expanded ontological-epistemological significance: it has always been multimodal, historical, diverse, plural, and contextually situated. It is intriguing that formal education has privileged normativity, failing to embrace language in its pluralities and diversities. For learners—of any age—being able to leverage multimodality to build communication and knowledge presents an exciting and expansive world in social interactions. This expansion is undeniably supported by advances in language and digital technology.

By participating in interactions and ontological-epistemological developments, Rodrigo Abrantes identifies the significant potential for constructing an education that embraces expansions such as collaboration, creativity, critical thinking, and the social perception of learners regarding learning, teaching, and re-signification. This also includes the environment and nature in which they live, fostering a setting where solidarity and empathy are practiced more. With this perspective, the researcher sees that schools can go beyond their traditional roles of qualification and socialization, offering the development of subjectivities and citizenship education.

The author's engagement with the academic milieu during their research proves to be fundamental in reaffirming the connection between schools and universities. The benefit is mutual for both parties, as is the inspiration. For instance, in observing the tripartite relationship of school, subject, and society, questions may arise that warrant investigation regarding this relationship, as well as what should be researched—or deemed appropriate—within the academic sphere itself. Schools and universities have been highlighting social changes, indicating shifts in students, teachers, families, occupations, languages, relationships, values, entertainment, and what else? The results of these studies, in turn, should be disseminated to schools, universities, and citizens alike. Within universities, new research and novel/revisited knowledge can circulate with greater fluidity and frequency, although this is not always the case. A closer relationship between schools and universities can only be beneficial for all parties involved.

I would like to highlight the innovative achievements in this book:

(1) The author recognition of the potential to work educationally with a non-commercial platform. At some point in the present-future, through non-commercial platforms (where commercial data transactions are not predominant), a differentiated large-scale education could be promoted. What kind of education would this be? One that seeks conscious social coexistence among people, the environment, and a socio-cultural-economically less unequal society, with individuals more sensitive to the "human" character in social relations. In essence, more universities can/should or could/should construct platforms aimed at disseminating a differentiated educational proposal. There is a literacy project being proposed in various countries,

precisely rethinking the relationship between school, university, subjects, work, environment, ways of life, and societies, along with the discussion on real, virtual, fake, artificial, artificial intelligence, and generative AI.

(2) The work does not postulate the "importation of foreign models" to Brazil. It indicates the possibility that a country like Brazil, with its high-quality universities and researchers, could invest in its own platforms.

(3) The realization of the possibility of expanded ontoepistemological work in Brazilian Basic and University Education. Learning through peers, where there is interaction between them, enables the expansion of the vision about "constructing knowledge" and the various forms for doing so. Rodrigo Abrantes' research opens up the discussion about alternatives in learning and teaching through cyber-social systems.

(4) A revisitation of educational proposals undertaken by elementary and secondary schools, as well as universities.

(5) A reexamination of teacher and student training, as well as civic education.

Regarding items 4 and 5, I emphasize that the research conducted by the author of this work was built from the perspective of Literacies Projects (as educational initiatives in current movements focused on educational revisions are called), which revisit (a) education; (b) the strong presence of coloniality in education and social formation; (c) educational–social–cultural principles on learning to read and write, being, constructing knowledge, making meaning, building languages, educating, communicating, interacting, relating, integrating, and solidarizing; (d) the relationship between person, citizenship, environment, ancestry, work, and future; (e) the relationship between self, selves, us, other, and others; and much more. Rodrigo Abrantes da Silva's background as a historian, combined with his extensive knowledge in technology, certainly enabled the broad scope of the research.

The author's research introduces a novel discipline (perspective) in digital literacy studies, making a significant impact on the current theoretical landscape. Moreover, it has succeeded. Remarkable!

Readers already know what to expect in this book: contributions to the field of education, precisely within a perspective that facilitates revisions, integrating experiences, experiments, and knowledge generated by the digital realm. Indeed, this book is concerned with the "formation of social agents capable of facing, individually and collectively, with science and humanity, the multiple crises of our times," a formation that can "provoke students to make sense of the world and then assume responsibility for acting in it," as can be verified. It presents itself as an essential and impactful reading for the educational rethinking of the coexistence between digital society and the society of writing.

Walkyria Monte Mor
Associate Professor/Tenured Professor
FFLCH—University of São Paulo
Co-founder of the Literacy Project:
Co-founder of the Literacy Project. DGP/CNPq-USP
wmm@usp.br

In *Cyber-Social Environments in Education*, Rodrigo Abrantes gives us a glimpse of the future of education. In a world beset by inequalities, including access to digital tools and resources, he shows how low-cost and web-accessible tools can open new opportunities for learners. This book is both theoretically inspired and empirically grounded. Theoretically, Abrantes articulates a version for cyber-social learning where the cyber is firmly in the hands of the social. In this frame of reference, the machine facilitates the social by supporting peer collaboration and collective intelligence.

Empirically, the book captures the rich and complex experience of learners who have become active makers of socially engaged knowledge. Rooted in the specific challenges of the Brazilian experience, the implications of this work are, nevertheless, global. Today, no place in the world is exempt from the challenges and opportunities arising at the intersection of educational inequality and digital media. This book tentatively points us all to some ways forward.

Bill Cope/Mary Kalantzis
Professors in the Department of Educational Policy, Organization, and Leadership at the University of Illinois Urbana-Champaign (USA). Dr. Bill Cope is also affiliated with the Information Trust Institute and the Health Care Engineering Systems Center at the University of Illinois Urbana-Champaign's College of Engineering. Dr. Cope and Dr. Kalantzis developed the educational platform CGScholar and are responsible for its management.

ABOUT THE AUTHOR

Rodrigo Abrantes da Silva holds a PhD in digital literacies from the University of São Paulo (USP). When pursuing his doctorate, he served as a teaching assistant in the Learning Design and Leadership Program at the University of Illinois Urbana-Champaign (UIUC). A historian graduated from USP, he has teaching experience in both K–12 education and higher education, as well as mentoring teachers in technology-driven learning practices. He currently works as an educational technology tutor and teacher at Colégio Dante Alighieri in São Paulo. He has coauthored books and published articles, for which he has received international awards. For more information, please visit: rodrigoabs.com.br.

CHAPTER 1

The Technological "Tsunami" and Education

Talk, talk, bet, as long as you talk, right?
Here is the box from which all the gifts of language emerge, a Pandora's box.
Jacques Lacan[1]

We are living through a technological revolution in constant acceleration. Drawing on Lacan's metaphor of Pandora's box in its digital correlate, we never fully know what might emerge from this kind of perpetual motion machine that characterizes current computational language models, from which generative artificial intelligence (GenAI) programs have become capable of writing and teaching themselves. The applications of digital technology are vast, expanding every day with each feedback loop; they permeate our daily lives through the ubiquitous presence of computers, cell phones, virtual assistants, smart cameras, and more.

The era of the digital society has arrived, even though socioeconomic inequalities in accessing technological resources still need to be overcome. Platforms and social networks have reshaped how people connect and share their lives, for better or worse. AI is now widely integrated into computer and smartphone operating systems, platforms, and messaging apps, while autonomous vehicles are being designed to revolutionize the transportation industry. Content streaming services have transformed the way we produce and consume films and music. In the financial sector, blockchain technology has introduced cryptocurrencies, paving the way for changes in financial transactions. The Internet of Things (IoT), now present in a multitude of devices, collects data from the physical world and converts them into information that drives interactions between humans, machines, and the environment, while 3D printing is paving the way for personalized object reproduction.

[1.] Lacan (2007).

In education, online learning platforms such as Coursera, edX, and Khan Academy provide a vast array of courses and educational resources accessible to anyone with internet access. Educational institutions use learning management systems like Moodle and Blackboard to provide course materials, assign activities, and utilize online communication tools. Additionally, virtual reality (VR) and augmented reality (AR) are beginning to be employed to create immersive learning experiences. AI powers intelligent tutoring systems, content personalization, and data analysis. Every day, thousands of people use YouTube to access instructional content on a wide range of topics.

However, in many educational institutions, technology is still predominantly used merely as a new facade for conventional pedagogical practices, based on the linear transmission of knowledge, assessment tests, and content mastery by the student. Formal education has not kept pace with the possibilities for interaction and transformation generated by the technological revolution, while also reproducing one of the problems of digital society, which has come to value results expressed in technical language and, as a result, has failed to grasp the essential aspect of our relationship with technology, which lies in human values.

We will discuss the issue of values and culture, which should later serve as fundamental premises for education. What is important to highlight at this moment is the trend of applying tools characterized by hypertext and network design, among other functionalities that have become so well integrated into people's daily lives, within a linear learning relationship. This approach does not meet the needs for the full development of knowledge, culture, critical thinking, and specialization required by digital society.

Technological acceleration generates unprecedented social dynamics, through which many traditional institutions and forms of knowledge lose their effectiveness, as they continue to reproduce conventional pedagogical processes and the corresponding classroom discourse—teacher speaks, student responds, teacher evaluates. In this dynamic, the prevailing conception of learning is the transfer of knowledge from the teacher's mind to the student's mind. Although this approach has been challenged over the past decade by proposals for blended learning and active methodologies, it continues to be used in many classrooms.[2]

However, in the digital society, students are expected to access constantly evolving knowledge and to appropriate that which is relevant to their life contexts, in a

---

[2.] We refer here to the concept of "conventional classroom discourse," as proposed by educator Courtney Cazden, a professor at the Harvard Graduate School of Education and a member of the New London Group, in her book titled *Classroom Discourse: The Language of Teaching and Learning.* With this concept, she aims to describe and analyze a model that, in some contexts, is termed didactic teaching and, in others, transmissive teaching.

process that involves critical re-elaboration and the creation of new knowledge. This dynamic can be described as a "design process" and involves behaviors, habits, skills, and ways of acting and thinking in the digital society—topics that will be analyzed throughout this book.

From Chalk to Chip: The Technological Revolution in Education

As a high school history teacher, I began incorporating digital technology into my classes to make them more engaging and interactive for students. This practice started around 2010, when I noticed most students began bringing internet-connected cell phones into the classroom and I saw an opportunity to use languages familiar to them. This started in 2010, when most students began bringing internet-connected cell phones to the classroom. Each student had in their hands a "cognitive prosthesis,"[3] capable of accessing thousands of knowledge bases in a matter of seconds. I aimed to leverage this resource in their learning process. I incorporated blogs, social networks, apps, and games into their daily repertoire of reading, writing, and interaction.

The activities generated astonishment and skepticism among my peers, who were used to a practice that prohibited the literacies students brought from outside the school. However, I continued with this nascent model of "blended learning" and began supporting colleagues interested in this approach, assisting them in creating activities with digital resources for their respective fields of knowledge.

I started this work at a private school in a middle-class neighborhood in São Paulo and later worked on teacher training projects in different states across Brazil. Many educators resisted the use of digital technology. At that time, most teachers could not have imagined that, 10 years later, knowing how to work with digital technology would become a prerequisite for continuing to teach during the long months when schools were closed due to the COVID-19 (coronavirus disease 2019) pandemic, from 2020 to 2022.

In recent years, I have observed an increase in educators' engagement in developing digital solutions, both in startups and in large technology companies. Based on my experience with some of these companies, I can affirm that, generally, teachers are seen as important interlocutors. They not only help understand the educational context but also bring creativity and a humanized perspective to

---

[3.] The term used by Cope and Kalantzis (2009) for the use of digital devices such as cell phones.

the workplace. However, not all teachers are able or willing to associate with the educational technology industry. According to my experience, this is because the interaction between this sector and schools and universities is often marked by conflicts between market interests and pedagogical principles. Nevertheless, I consider it positive that teachers occupy these spaces and influence businesses and projects involving the use of digital technology in schools.

The public version of ChatGPT released in November 2022 was based on OpenAI's GPT-3.5 model (OpenAI, 2022). Once again sparked debates among educators, ranging from apprehension to excitement. To gain a more balanced perspective on such a disruptive technology, I suggest adopting a paradigm that allows for its integration, recognizing its potential while also analyzing it critically, especially in the context of the teaching and learning process. However, this endeavor is by no means simple; on the contrary, it is marked by significant complexity and requires certain dispositions to incorporate, among the many digital technological resources, a machine capable of writing based on a vast volume of texts.

In this book, I present an exercise in the complexity of integrating learning experiences in a digital environment. It is drawn from my doctoral study in digital literacies at the University of São Paulo (USP), where I sought to demonstrate that it is possible to enhance learning practices driven by technology, enabling students to integrate their knowledge processes and personal experiences into the creation of knowledge. In other words, I advocate the perspective that digital technologies can be used in a democratic, inclusive, and participatory manner in education, expanding practices already explored in the field of literacies in Brazil. The application of my research study is described in Chapters 2 through 5, which present reports and analysis of experiences with an educational platform, CGScholar.

Before we delve into the practical aspects, let us explore further the topic of the impacts of digital technology on conventional education systems.

The Cyber-Social Perspective

The advent of digital technology has triggered changes as profound in society as those caused by the emergence of writing. Just like writing, digital technology is a human invention that transforms our ways of thinking and organizing society. While oral language develops from the human body, writing and digital technology are external elements that have revolutionized the way we communicate and store knowledge.

The literature extensively documents that writing profoundly transformed the ways of thinking and organization in oral cultures. It developed the power to modify how we think, in a complex interaction between language and thought. One example of this process is the shift from dialogue-based intellectual practices, rhetoric, and speech to written text, which began to be organized within the visual structure of a page. In the modern world, this trend was accelerated by the printing press and literacy education, which used text as a central tool for mass education.[4]

Furthermore, due to the characteristics of written language predominant during this period, there emerged a need to standardize and simplify linguistic diversity. This standardization was deemed essential to facilitate didactic education, as well as to enable the evaluation and control of the learning process. Monte Mor describes this moment as follows:

> *In this didactic approach, meanings are also standardized and homogenized; agreements are made to create established and official versions of standard languages; the social functions of writing become systematized, thereby reflecting the needs of dominant communities; in short, the privilege of written words is preserved. In this proposal, the pursuit of homogeneity becomes evident when contrasted with the previous moment in the history of language. (Monte Mor, 2017, p. 273)*

In contemporary society, computers provide a new way to materially produce meanings. However, I do not intend to assert whether this is good or bad but rather acknowledge the existence of new phenomena that need to be studied. In this context, a culture emerges that operates in an expanded manner, through various means of communication, as well as diverse and stimulated forms of interactivity. There is a perception of other possibilities for reading, interacting, seeing, thinking, and engaging with choices or situations that differ from conventional patterns.

The most representative creation of this new era is the internet. For this discussion, I will focus on the emergence of the World Wide Web (WWW), which enabled the reception, creation, publication, sharing, and reading of documents

---

[4.] Graff (1987) *apud* Cope and Kalantzis (2023a); Street (1984) *apud* Cope and Kalantzis (2023a).

containing text, images, videos, and sounds, thus expanding the possibilities for representing and communicating information and knowledge. Also known as Web 2.0, this phase of the internet is characterized by user-generated content across various platforms. The ways of participating in digital culture have been expanded, allowing people to use the Web's infrastructure to engage in multiple and varied affinity spaces.[5]

One of the reflections of the transformations in the cultural context of the internet, according to authors such as Cope et al. (2011), is the significant shift in the location and recognition of epistemic authority.[6] They say that this authority is now more flexible, subject to specific conditions, and may be temporary, generally based on claims of possibility ("may") rather than certainty ("is"). The authors argue that these claims are more likely to be questioned and critically analyzed, based on personal experience and individual voice.

This culture of participation has generated new literacies and forms of association that question many of the foundations of modern social organizations. For example, instead of vertical hierarchies, what Castells called a "network society" emerges. In place of the printed page, there is the interactive, multimodal, and hypertextual page. However, to fully thrive in this new world, and not merely reproduce the old one, it is necessary to adopt a new mindset, a mode of interpretation that aligns with practice.

Cope and Kalantzis argue that it is necessary to name this new regime as cyber-social, given the scale of the changes underway:

> If a regime now passing announced itself under the banner of literacy, we're going to name our regime change 'cyber-social meaning,' where computers have so completely inveigled themselves into the vernacular meanings of our lives that they have in some practical ways changed our human natures. (Cope & Kalantzis, 2023a, p. 2)

According to the authors, although the changes triggered by digital media already have a few decades of history, we are only at the beginning of the transformations. They note that the first applications of computers for human vernacular meaning were developed in the context of formal education. The

---

[5.] A concept proposed by sociolinguist James Paul Gee (2015) to describe environments where people with common interests and goals come together to share knowledge and skills. Chapter 2 provides a brief description of the development of the Web.

[6.] Por autoridade epistêmica, refiro-me às instâncias—pessoas e/ou máquinas—que enunciam conhecimentos passíveis de serem reconhecidos como saber válido.

PLATO[7] project for teaching and learning, the first educational and multimodal computer, was launched in 1959 and demonstrated the dynamics and applications of social media.

> *Nevertheless, we are still in a period that Jean-Claude Guédon has called the digital incunabula, the Latin word for 'cradle' that bibliophiles apply to the first fifty years of the printing press. Despite the rapid diffusion of print technology after 1450, its characteristic textual apparatuses of contents pages, indexes, bibliographies, and page numbers did not appear until the sixteenth century (Eisenstein 1979; Guédon 2014). In our own time, the key features of cyber-social meaning may only now be coming into view, more than seven decades after Turing's Manchester Mark 1 computing machine and six decades after PLATO's combination in a computing machine of text, pixelated image, sound, and networked connection of persons. (Cope & Kalantzis, 2023a, p. 2)*

In terms of historical time, digitality is a very recent process, and we have only two points of comparison to sketch some analyses: the era when writing emerged and the advent of the printing press. The parallel proposed by Cope and Kalantzis in the earlier citation suggests a similar latency period between the emergence of printing technology until the definition of the elements of the printed page and the emergence of the computer and the current form of web pages. With all due proportions considered, I would say that several more decades will be necessary before this regime finds a new balance. A task to be undertaken to achieve this new balance is to recognize the new regime and name it. Therefore, I will elaborate more on the cyber-social era throughout this book.

Cybernetics: The Science of Feedback

The prefix "cyber," as I use it here, originates from the word "cybernetics," an interdisciplinary field of study that emerged from the work of mathematician Norbert Wiener on the regulation of systems composed of humans and machines, based on feedback mechanisms. The term first appeared in 1948 with the publication of his book *Cybernetics: Or Control and Communication in the*

[7.] PLATO (*Programmed Logic for Automatic Teaching Operations*), também conhecido como Projeto PLATO, foi desenvolvido na Universidade de Illinois.

Animal and the Machine. Wiener derived it from the Greek word "kubernētēs," which shares its etymological root with "government" and means the helmsman of a ship (Wiener, 1961). This metaphor illustrates a basic cybernetic system, composed of a machine (the ship) and humans (rowers and helmsman), with the open sea as its environment. The helmsman has a purpose—to steer the ship in a specific direction—which requires a keen awareness of environmental disturbances (currents and winds) and, based on this information, the ability to communicate the necessary movements to the rowers to maintain the course. These adjustments occur through feedback loops, resulting from a dynamic and adaptive interaction between the system and its surroundings.

In the cyber-social perspective, the term "cyber" does not refer solely to technology:

> *Incidentally, by 'cyber', we don't (just) mean technology. To return to the creator of the 'cyber' metaphor, Norbert Wiener, cyber is not about technology per se—as illustrated by the kubernētēs or steersman on the ancient Greek ship (Wiener 1948/1961). It is about feedback relationships that may be between humans and machines as much as relationships within more or less intelligent machines. Beyond computer-mediation and machines, 'cyber' turns on the iterative, recursive relationships that can motivate social action and drive forward designs for social change. (Tzirides et al., 2023, p. 99)*

Cybernetics has become the science of feedback, studying systems composed of humans and machines and the recursive relationships they develop over time. Cope and Kalantzis draw on this tradition to question the view of AI as a reproduction of human intelligence, which leads people to perceive AI as if it were human. Metaphors such as "neural networks" and "deep learning" reinforce this view by suggesting comparisons between human attributes and machine behavior.

What is the consequence of this? By equating the brain and human behavior with the functioning of computers, we end up viewing humans as machines and vice versa. Brazilian neuroscientist Miguel Nicolelis (2020) argues that comparing the human brain to the functioning of an algorithm can lead to a gradual remodeling of the process by which our brain acquires, stores, processes, and manipulates information. Cope and Kalantzis (2023b) note that this logic also seems to underpin the rise of behaviorist approaches in the field of computer-mediated human learning.

However, for cybernetics, "cyber" and "social" have always been two distinct forms of intelligence that can complement each other. From this perspective, the authors propose the term "cyber-social" to replace "artificial" with "cyber" and, with "social," to emphasize the human element in the relationship with technology (Cope & Kalantzis, 2023b, p. 19). A similar effort has been undertaken by researchers across various fields (Hui, 2024), who have turned back to cybernetics to envision technological approaches that transcend both the monopoly of platforms and their consequences, as well as some transhumanist visions that feed into dystopias of immortality and superintelligence.[8]

In my doctoral study, instead of thinking of AI as a reproduction of human intelligence, I proposed that we position ourselves in a cyber-social relationship with machines, that is, one of complementarity. In other words, the research sought to converge with Hui's (2024) view, which argues that we need to reimagine our relationship with technology to overcome the way it is embodied at the beginning of the 21st century. For this effort of (re)imagination, the cyber-social perspective offers some important insights. According to Cope and Kalantzis,

> The intelligence of humans is much more than their bodies plus brains. Their meanings are also outside their bodies—in their materialized experiences, life histories, physical environments and social settings. The meanings of our intelligence extend beyond our bodies—in the way our kitchens are laid out, the way we handle tools, the ways we present ourselves to others and see ourselves in their seeing of us when our bodies pass, and in every new social encounter what we have learned from the loves and hates we have had—a whole lot of stuff that is around us, not just in us. (Cope & Kalantzis, 2023b, p. 13)

In this view, lived experiences incorporate meanings into the bodies, which have the ability to project learned representations onto the external world through interaction in social environments. This perspective understands that machines expand the capacity of natural language to name the world and construct meanings, becoming part of the human experience rather than replacing or reproducing it.

Through this perspective, we understand that consciousness cannot be reproduced, as it is not computable, according to the neuroscientist Kristof Koch. It is

---

[8.] I refer to proposals such as the fusion between humans and technology, as advocated by American futurist Ray Kurzweil (2000, 2015), and the mind behind projects like Google's Calico, which aims to overcome death, as dystopian. Information about Calico's projects can be found at https://www.calicolabs.com/, accessed on August 24, 2024.

the bio-physical presence in the world and the experience of affects, sensations, and ideas that give meaning to life—"the sense of one's own life."[9] Therefore, I believe that the cyber-social approach addresses the complexity involved in the current relationship between humans and machines, human intelligence and AI, and the ongoing transformations, in such a way that we can evaluate their impacts on language and literacies processes in education.

The Expansion of Literacy into the Digital Realm

I take as an example some repertoires of characters used to express meanings. The common characters found on keyboards and in mobile device emojis are human creations. These characters have been developed over time to meet the needs of communication and visual representation. Today, the available repertoires of characters are remarkably extensive and diverse. This can be observed in translingual script projects like *The Noun* and *Unicode*,[10] which offer repositories of characters, images, icons, and emojis that can be utilized across various languages and communication contexts. These resources provide a wide range of expressive possibilities, enabling diverse communication of meanings through characters and symbols.

The cyber-social perspective highlights the consequences of the mechanization process involved in the transposition of forms of meaning. This mechanization is fundamental to the interoperability of digital devices, meaning that the functionalities of our computers and smartphones are global and not subject to national or regional criteria. The conversion of meanings into binary notation

---

[9.] Koch (2019) *apud* Cope and Kalantzis (2023b, p. 12).

[10.] Unicode is a character encoding standard designed to represent text consistently and comprehensively, regardless of language or computer system used. Currently, it includes 149,000 characters, or graphemes. It assigns a unique number, known as a Unicode code point, to each character, symbol, or ideogram used by all known languages. The primary goal of Unicode is to overcome the limitations of traditional character sets, which were often specific to certain languages or geographic regions. Unicode enables different systems and platforms to communicate and display text coherently, avoiding interoperability issues between different character sets. It covers a wide range of characters, including alphabets, ideograms, mathematical symbols, emojis, and more. The Unicode system is maintained by the Unicode Consortium, a non-profit organization, and is widely adopted in the software industry, on the Web, and in text encoding standards worldwide. Unicode's popularity is evident in its widespread use in programming languages, operating systems, web browsers, and many other software applications. The Noun Project is an organization dedicated to building a global visual language. This visual language aims to enable quick and easy communication, regardless of who you are or where you are. The Noun Project boasts the most comprehensive and diverse collection of icons in the world. With a community of designers from over 120 countries, they are creating the most diverse and extensive iconography collection ever made.

makes it possible to encode and decode proper names in any language, rendering cyber-social meanings a translingual phenomenon (Cope & Kalantzis, 2023a).

In the cyber-social regime, natural language is supplemented by non-natural mechanical language, resulting in an extraordinary expansion of the expressive capacities of language. In this context, the Web can be seen as a form of social mind, a space for interaction and communication where enormous repositories of human experience are available. This transformation presents a magnitude unparalleled by typographic literacy. To navigate and inhabit this new world, humans rely on cognitive prostheses—such as cell phones, computers, and IoT devices—that enable them to access this vast collective memory and apply it in everyday social practices. This dynamic surpasses the limitations of individual human memory.

Instead of trying to store all the content of the Web in our natural memory, we use digital devices as memory assistants. These devices not only expand our storage capacity but also facilitate the retrieval of information, which is essential for our continuous interaction with the digital world.

Given the ways in which technology has expanded our ability to manage knowledge and construct meaning, it becomes necessary to reconsider conventional education models. Reflecting on this, Monte Mor (2017) argues that young people should no longer be seen as "the same students" as in previous decades. There is a new generation. Following Prensky (2010, 2012), she states:

> *Young people express having higher expectations of school than just conventional classes and content they consider disconnected from their real contexts. They expect to be respected in their positions and constructions of meaning, in their interests; they would like to create more, see greater connection between school and their environments. [...] In 21st-century education, the problems of tomorrow cannot be solved with the minds of yesterday. (Monte Mor, 2017, p. 277)*

In the author's view, the process of school literacy should keep pace with societal changes, reflecting transformations in language, relationships, work, and the ways these elements are constructed. However, this necessary transformation conflicts with conventional approaches. When these perspectives are reproduced in the digital environment, they create a post-typographic equivalent of the functional literacy of the printed world, simplifying the complex language practices of collaborative and participatory cultural contexts and reducing them to mere sets of operational techniques.

Authors such as Lankshear and Knobel have captured this phenomenon in traditional and hegemonic definitions of digital literacy. They observed that these definitions converge toward an engagement with information:

> *Digital literacy is constructed in what we might call 'thruthcentric' ways, and as some kind of defence against being manipulated, improperly persuaded, or duped. It is invested with values and orientations associated with liberal and 'critical' conceptions of media awareness and the like. (Lankshear & Knobel, 2015, p. 12)*

It is not to say that media literacy criticism is not important, but to point out that, in terms of literacy practices, this conception does not capture the complexity of textual interactions involving communication and relationships on the Web. By questioning the excessive emphasis on the truth and credibility of information, the authors argue that social practices on the internet often focus on relationships and affiliations, where people seek to join certain groups for reasons of identity and recognition. Finally, they criticize the idea that digital literacy (in the singular) is a single competency, arguing that it represents a variety of social practices and conceptions of meaning mediated by digital texts. Therefore, digital literacy should be seen as a set of "digital literacies" in diverse and multifaceted contexts.

For example, these authors demonstrate that digital literacies are heterogeneous in daily life: even blog posts—blogging or fanfiction—represent highly diverse digital practices. Blogs, originally appearing as lists of hyperlinks to other sites, now range from personal diaries to media and commerce critiques. Similarly, fanfiction involves a wide variety of narratives created by fans of television shows, movies, and other media. These practices challenge the idea that digital literacy is a single competency and highlight the plurality of digital literacies in diverse contexts.

People engage in literacy practices based on specific cultural insertions, critical and creative ways of doing things, rather than operational techniques. Successful blogs, for example, are more the result of the author's perspective and style than of operational management. Furthermore, Lankshear and Knobel reveal that young people participating in online gaming communities view digital competence as a matter of exploration and experimentation, rather than merely following technology usage rules. This reinforces the idea that digital literacies are built more on cultural practices and ways of creating meaning than on operational skills.

Based on the studies of Lankshear and Knobel, Monte Mor proposes that education should focus on the concept of performance epistemology, according to which:

> *learners construct knowledge independently of having learned examples and models on how to build it. To do so, they use intuitive or non-intuitive learning, developed inside or outside schools, such as assembling, bricolage, collaging, editing, data and information processing (and others), which transform into constructions or reconstructions that meet their plans or endeavors. (Monte Mor, 2017, p. 279)*

A pedagogy inspired by such views can provide teachers with the environment and disposition necessary to engage young people in meaningful learning with implications for the world of life. This becomes even more important considering the speed of transformations resulting from the technological revolution.

CHAPTER 2

The Making of a Cyber-Social Learning Environment

Today, the traditional and careful construction of the human being separate from the rest of the world is collapsing. I imagine this process as the fall of a giant, decayed tree. It does not cease to exist, but its function changes. It transforms into a space of even more intense life, with the germination of other plants, colonization by fungi and saprophytes, and occupation by insects and other animals. In fact, the tree itself is reborn from its seeds and roots.

(Olga Tokarczuk, 2023)[1]

Writer Olga Tokarczuk reflects, in the above passage, on the transformations of the contemporary world from a perspective that believes a major transformation in the pillars of society is necessary. I believe that educational systems represent a fundamental part of this transformation. By applying the writer's metaphor to educational systems in the digital society, I observe, with some optimism, that the process experienced by the tree signals possibilities for rebirth, which can occur through windows of opportunities (*affordances*)[2] provided by digital technology.

Expanding this reflection to a broader understanding of social meaning-making systems, I observe that the literacy regime, dominant in recent centuries, is also undergoing a phase of expansion and renewal. Findings from my study suggest that this renewal may occur in certain Brazilian educational contexts. Thus, I aim to provide in this book insights and perspectives for teachers who wish to engage with the ongoing transformations.

In this chapter, I propose a characterization of the cyber-social perspective, starting from the reconstruction of historical aspects of one of its fundamental

---

[1] Translated into English by me, from the Portuguese edition (*Escrever é muito perigoso: Ensaios e conferências*) published by Todavia in 2019, pp. 18–19.

[2] This is a concept that will be revisited later and will play a significant role in analyzing the learning opportunities generated by the cyber-social environment.

pillars, the World Wide Web.[3] I also present proposals aimed at highlighting the expanded possibilities for human action within the cyber-social context, including the pedagogical use of machine-mediated peer review and the utilization of artificial intelligence (AI)-based learning analytics tools as a complement to the human learning process. These methodologies will be discussed in Chapters 4 and 5, respectively.

In Chapter 2, I will discuss the relevant characteristics of the Web in learning situations and about affordances. However, before that, I will review some of the tensions present in the contemporary educational debate.

Rethinking Education: Toward a Humanized Approach to Technology

This book argues that the use of digital literacies in formal education can open up possibilities for transformation within educational institutions.[4] However, in recent years, the debate around the impact of technology on education has increasingly focused on the use of data by big tech companies. This is an essential topic and has been addressed with great competence by various fields of knowledge, as well as by established authorities, although somewhat belatedly.

Nevertheless, the focus mentioned earlier is not pedagogical; it does not address the possibilities of using digital technology for learning, but rather assesses the political implications of the private monopoly over the data of billions of people. It is certain that educational institutions need to respond to this scenario and define their policies with discernment, as these will shape what can be done in the classroom in terms of technology. Nevertheless, these issues are more political

---

[3.] The Web, here, refers to the part of the internet that most people use to communicate with friends and family, perform daily tasks such as financial transactions and shopping, and engage in affinity spaces and learning environments. I am not referring to the entire global network of computers connected through TCP/IP (Transmission Control Protocol/Internet Protocol), which constitutes the internet. The Web originated through a type of TCP/IP used to transmit data between a browser and a server, called HTTP (Hypertext Transfer Protocol), allowing pages to contain links to other documents.

[4.] Literacy studies differentiate between the term "digital literacy" in the singular and "digital literacies" in the plural. The singular form refers to a model equivalent to the functional literacy of typographic literacy. This model works with the idea of a "digital literacy" that can be transmitted in the same way to everyone and focuses on technical operations, such as software handling. On the other hand, the plural form indicates that there is no single universal literacy but rather different cultural ways of using digital technologies in various contexts. This perspective emphasizes that literacies emerge from social practices and warns against the risk of abstracting them from their social conditions to turn them into abstract concepts. In this study, the concept of "digital literacies" is understood as enabling students to use technological resources to develop an active, critical, and creative approach to the knowledge presented by the teacher.

than pedagogical, and my aim here is to evaluate the implications of using digital resources in the relationship between teachers and students.

Another field that also presents a strong discourse on technology is mental health, associating technology use with negative impacts such as loss of cognitive abilities and the disintegration of social bonds. The solutions proposed by this field converge on reducing "screen time," but this measure does little to humanize people's relationship with technology and can also generate moral panic and distrust. In the educational field, the pandemic experience and remote learning intensified these debates, with critics arguing that excessive digitalization can disintegrate public education and the essential human relationships necessary for learning. The concern that technologies should not overshadow the human element is valid, but there is also a lack of a clear guideline on how to integrate digital tools into education in a humanized way—something I sought to address in my study by introducing the cyber-social approach (Nóvoa & Alvim, 2020).

Digital Education: From Conventional to Cyber-Social

The challenges highlighted earlier can be debated and addressed from other perspectives. In education, it is well known that people often think about and use computers with mental models that were not developed to interact with them.[5] In other words, we carry a conventional ethos that, when faced with innovative technology, ends up merely reproducing a diminished conventional ethos in the digital realm. As a result, and based on the key authors considered during my study, I believe that to align education with the characteristics of digital society, it is necessary to recognize the subjective effects that technology has on students and to work with new epistemologies[6]—that enable people to expand their capacity to learn.

In this perspective, learning is no longer viewed as something that occurs solely within the individual but as something that involves the individual as a member of a collective. Memory shifts from being individual long-term memory to becoming collective memory and social mind. In this way, technology becomes an indispensable part of people's humanity rather than a negation of it. Students are then seen as active learners, knowledge creators, flexible and collaborative,

---

[5.] Lankshear and Knobel (2011), Monte Mor (2017), Gee (2013).

[6.] The term epistemology is related to the Theory of Knowledge, but here it can be understood as the ways of knowing, thinking, and actions taken to generate knowledge about the world.

capable of applying divergent and insightful thinking in constantly changing contexts, and willing to take risks in innovative and creative ways.

Given this, I firmly believe that it is essential to educate students to become determined learners, capable of continuing to learn throughout their lives—a crucial skill in digital society. Moreover, this approach prompts us to reflect on the need to modernize educational systems, rethinking digital education models at all levels.

The cyber-social learning approach provides an accessible opportunity for students to acquire the skills needed to create knowledge relevant to the social practices in which they are involved. This helps to avoid the risk of rigid knowledge that is disconnected from the needs and desires of individuals, where a single agent holds authority over what should be known. In this way, everyone can develop knowledge that is important to their life and work contexts.

To put this process into practice, we need to free ourselves from the presumption that the teacher knows everything students need to know and do. For only if the teacher could become each of their students and experience their life conditions could they have such a perspective. It is not easy to work with the discourses that students can produce. Here, I think in terms of agency, moments of representing ideas, and how these can circulate in the classroom, contributing to formal learning.

I am aware of the challenges of working from this perspective given the current state of education in Brazil, but I assert that it is necessary if we want to provide our students with better chances of success in the digital society. As I will show in Chapter 2, I believe it is important to work with students' knowledge not only for recognition but also as a valuable learning resource for the classroom and beyond. I understand that this approach is highly conducive to learning—a type of learning that can lead to the production of knowledge necessary and relevant to local contexts.

The cyber-social paradigm aims to foster this dynamic. Within this framework, the concept of staying updated takes on a significant breadth: people will need to know how to update their Knowledge skills in relation to their contexts, relying not solely on knowledge provided by educational institutions but on something each individual will seek to complement with their own knowledge and circumstances. In this type of relationship, educational institutions will play a vital role by providing the dynamics for knowledge construction, promoting a regime of collaborative intelligence, and offering technologies that can expand these capacities.

New Subjectivities in the Digital Age:
Interactions Between Humans and Machines

The Digital Age has brought about the possibility of new subjectivities that can enrich cultural experience and our references by expanding communication and interaction through the integration of human and machine languages. While these impacts can also generate adverse effects on identity formation, empathy, critical thinking, and mental health, this book focuses on highlighting opportunities to maximize the best aspects of this new reality within a digital education environment.

In the cyber-social environment, we combine human-built knowledge with machine capabilities. Integration between human language and machine language—through tools like automatic translation and virtual assistants—connects us with people from different cultures and languages, overcoming linguistic barriers, facilitating the naming and categorization of complex concepts, and improving our ability to organize and access information. Online platforms democratize access to educational resources, promoting continuous and collaborative learning that enriches our perspectives and skills. The internet offers us the opportunity to learn from anywhere, at any time, and to contribute to the advancement of global knowledge.

If we understand that the mechanical language of computers expands the naming capacity of human natural language, it becomes easier to see the advantages of complementarity between humans and machines. The Web is an example of a machine that expands human language. Its communicative potential can be better utilized in formal education, based on my observations and what the literature has reported (Cope & Kalantzis 2013, 2017; Lanksherar & Knobel 2011). Teachers can use it to more easily implement project-based and/or community-based learning proposals, develop lifelong learning habits, explore new forms of creativity and innovation, and better meet the demands for more meaningful participation in the digital society.

The evolution of Web language through HTML5[7] improved resources for handling alphabetic text and images within the same interface, while text markup tools made it possible to translate discourses into any language cataloged in

---

[7.] HTML5 is the fifth version of the hypertext markup language (HTML), used to structure and present content on the Web. This version represents an advancement toward a more open and dynamic web, providing better support for multimedia, graphics, and interactions.

the Unicode system. This condition made multimodality and multilingualism a reality in web literacy practices.

Thus, the Web has triggered changes in three areas related to education: (1) it has added material conditions for producing meaning and expanded the possibilities for working with multimodal writing; (2) it has removed communication barriers related to differences between natural languages, although these barriers still persist in power relations; and (3) in formal education, it has expanded opportunities for listening to students, leading many teachers to seek to include them as active participants in the classroom.

Why Does the Linear Model of Learning Still Prevail in the Use of the Web?

Although it is already recognized that vertical hierarchies have changed on the Web, enabling horizontal relationships where the negotiation process between teachers and students is present, we still observe, as noted in Chapter 1, the predominance of classroom discourse organized by a logic in which the teacher delivers a lecture and asks questions, followed by some students responding, and then the teacher reflects, mediates, and evaluates.

This discourse is based on a flow of knowledge transmission from the teacher to students, a model that was established in the 6th century with the monasticism of Saint Benedict.[8] It remains a long-standing structure in society. The problem with having this hegemonic and predominant discourse in education is that it leaves little room for students to create meaning and diversify their learning strategies.

The transition of pedagogical practices to the Web, and specifically to *Learning Management Systems* (LMS),[9] has not significantly transformed learning relationships, as I present in Table 2.1.

[8.] Saint Benedict (c. 480–543), born in Italy after the fall of the Roman Empire, founded several monasteries and is credited as the founder of Christian monasticism in Western Catholic Europe. In *The Holy Rule of Saint Benedict*, he writes about the relationships between the abbot or superior and the monks within the monastery, explaining a mimetic pedagogy in line with the didactic teaching of a committed knowledge system: "For it is the master's role to speak and teach; it is the disciple's role to be silent and listen. Therefore, if something needs to be asked of the Superior, it should be done with all humility and respectful submission" (Kalantzis & Cope, 2020, pp. 54–56).

[9.] These systems are the foundation of online learning platforms.

Table 2.1 Comparison Between LMS Platforms and CGScholar

Platform	Google Classroom (Proprietary LMS)	Moodle (Open-Source LMS)	CGScholar (Sustainable)a
Key Features	Courses, management tools, quizzes, and assignments	Courses, management tools, quizzes, and assignments	Communities, publisher, bookstore, publication, analytics, generative AI integrations[b]
Pedagogy	Distribution of content and tasks	Distribution of content and tasks	Knowledge construction and publication
	Individual cognition	Individual cognition	Collaborative intelligence
Agency	Centralized, hierarchical	Centralized, hierarchical	Decentralized, heterarchical
Communication System	Web 1	Web 1	Web 3

[a]The CGScholar platform features a cyber-social design and was the basis for my doctoral study. Its history and functionalities are presented later in this chapter.
[b]These functionalities are described in Chapter 3.
Source: created by the author.

In this perspective, the new medium—the LMS—ends up further restricting learning by perpetuating pedagogical models created for societies where computers did not exist, resulting in even greater isolation than previous methods. This is evident in practices like using software that limits students to a single browser tab and in courses that promote student isolation at their computers, merely consuming content. Some criticisms directed at the current distance education (EaD [educação a distância]) system in Brazil, often accused of poor quality outcomes, are likely related to the lack of appropriate pedagogy and this type of digital medium.[10]

Furthermore, the main LMS platforms have barely tapped into the potential of the social Web (Web 2), let alone the emerging Web of big data, AI, and cyber-social meaning (Web 3). Although some of these systems, especially the proprietary ones, are currently adopting a data analysis discourse with a modern and futuristic facade, what they are actually proposing is merely the application of data analysis techniques to test results, establishing rankings and comparisons among students, schools, and education networks, and projecting these analyses into visualizations that may have visual appeal but bring nothing new to education. On the contrary, these analyses underpin intense competition at all levels of education, fueled by behaviorist procedures at the core: if the test is passed, there is a reward; if the test is failed, there is a punishment.

However, when we examine the communication infrastructure of the Web, we realize that it offers numerous opportunities for reading, writing, communication, and interaction, allowing knowledge to circulate more widely. The Web has introduced new communicative capacities, and when these capacities become available, people use them.

Because of this, the conventional classroom environment that relies solely on a transmissive approach can become demotivating, especially for young people who have grown up in the digital culture. I experienced the beginning of this digital culture in Brazil, with the arrival of the first video games, the internet, and email. During my childhood and adolescence, my access to books was limited. Living in a peripheral city with few cultural resources such as libraries, theaters,

---

[10.] In an editorial on November 9, 2023, the newspaper *O Estado de S. Paulo* criticized EaD because students from this teaching modality performed the worst in the National Student Performance Exam (Enade). The issue is not about criticizing the EaD currently practiced in Brazil, which indeed, based on my experience as a teacher in various courses within this modality, presents all the structural problems derived from the mere transposition of traditional pedagogies into the digital environment. The real issue is that the editorial fails to present any criteria for analyzing digital education and considers the current EaD as something natural, using only a comparison with in-person teaching as its sole evaluation metric. With such a limited analysis, the editorial can only conclude that "it is time to review the parameters for the education that the pandemic has spread." The pandemic could have been used as a window of opportunity to rethink and innovate in education, but the educational community entrenched itself even further in traditionalism. The editorial is available at: https://www.estadao.com.br/opiniao/graduacao-e-mais-do-que-um-diploma/, accessed on November 10, 2023. After the release of the Enade results, which measured the performance of higher education students in 2022, the Ministry of Education adopted a similar perspective as represented in the editorial and began to harshly criticize EaD, without presenting a better analysis criterion than the one in the editorial. In response to these positions, journalist João Vianney defended EaD by considering the aspects of social inclusion provided by this modality in the recent history of education in Brazil. His article is available at: https://www.cnnbrasil.com.br/forum-opiniao/mec-abre-guerra-contra-ead-sem-causa-determinada/, accessed on November 12, 2023.

and cinemas, school was my main connection to knowledge. Today, this has changed, except for those who are still deprived of basic resources due to poverty.

I would now like to summarize this discussion and propose some connections with education, considering the relationship between pedagogy, or the classroom, and the Web from the perspective of a communication system, and how this system aligns with its infrastructures (see Table 2.2).

Table 2.2 A Comparison Between the Characteristics of Web Phases and Pedagogical Models

	Web 1	Web 2	Web 3
Infrastructure	Content distribution stored on servers	Interactive and social	Cyber-social
Pedagogy	Conventional Transmissive Content distribution: • Video lessons • PDFs • e-books • Discussion forums • Tests • Assignments	Progressive • Hands-on • Dialogical • Students as co-creators of knowledge	Reflective • Hands-on • Dialogical • Students as co-creators of knowledge
Epistemology	The form of the media reproduces conventional epistemologies. It maintains its "constitutive" characteristics: hierarchical, linear, sequential, progressive-regressive (gradual).	The form of the media enables the expansion and diversification of epistemologies. It coexists with conventional epistemologies but allows their characteristics to	The form of the media enables the expansion and diversification of epistemologies. It coexists with conventional epistemologies but allows their characteristics to

		Web 1	Web 2	Web 3
			be broadened into heterarchi-cal,[a] non-linear, and non-sequen-tial forms.	be broadened into heterarchi-cal, non-linear, and non-sequential forms. Students bring their own reper-toires, and there are exchanges with a greater diversity of lan-guages and forms of expression.

[a]Regarding the concept of heterarchy, see Chapter 3, Kaufaman & Schunn (2011), Montebello et al. (2018), Keh (1990), Pinheiro (2020), Kalantzis and Cope (2016, p. 62).
Source: created by the author based on Berners-Lee et al. (2001), Hogan (2020), and Cope and Kalantzis (2023b).

Table 2.2 highlights how the discourse in the conventional classroom aligns with the structure of Web 1. Initially, my study aimed to understand why pedagogical practices in digital environments remained rooted in the logic of Web 1. I initially believed that the issue could be resolved by teaching teachers how to "operate" the software, but my early findings indicated that the root of the problem was not in the technology itself but rather in the learning relationships—in other words, it was more of a pedagogical issue than a technological one.

Society shapes educational institutions with their own language and culture, but this design is not transparent to people. Within these institutions, subjectivities are formed, and identities are represented and performed. This large socially constructed edifice does not appear as the result of planned and historically sit-uated actions, but as something "natural," that has always existed. Therefore, the challenges for transformative education are related to organizational processes, mindsets, and behaviors, as I will discuss in Chapter 3.

To expand conventional pedagogical processes, people need the opportunity to realize that there are possibilities for doing so. However, this realization does not happen overnight. It does not occur through short-term training or top-down directives. Given that digital technologies allow us to organize education differently (within the contexts of Web 2 and 3, as shown in Table 2.2), authors connected to the sociocultural perspective of literacies have developed theories that can help us recognize these possibilities and diversify our pedagogical practices, some of which I will present in the following.

Navigating the Digital Environment: *Affordances* and Their Implications in Education

In order to analyze the possibilities for action that digital technology provides to humans, I refer here to the concept of *affordance*, as proposed by psychologist James Gibson (1979/2014), based on an ecological conception of perception.[11] The ecological conception refers to the study of the relationship between an organism and its environment, characterized by reciprocity and complementarity. In other words, this view suggests that certain properties exist only through the interaction between the two and not independently. In this regard, it would already be possible to deepen a comparative analysis with the cyber-social perspective.

Affordance refers to the possibilities for action that an environment offers to an agent, emphasizing its potential and relational nature. This concept underscores that it is not the qualities or properties of the environment itself that are perceived but rather the opportunities for action it provides to the agent. The environment, therefore, offers resources and support that the agent can utilize to act, highlighting a relationship of reciprocity and complementarity. In essence, affordances represent what can be done, existing as characteristics of the environment that are specified in relation to the agent, regardless of whether they are perceived or not.

The concept of *affordance* can capture the complex relationship between humans and machines, highlighting their potential for complementarity, which becomes essential for certain actions to manifest. For these actions to occur, it is not enough for humans and machines to simply be present; both must behave in specific ways and interact in particular forms. From a cyber-social perspective, the possibilities for action emerge precisely from the recursive relationship between

---

[11.] According to Gibson (1979/2014), the verb "to afford" is found in the dictionary (meaning to provide, offer, or supply), but the noun "affordance" is not. Gibson gave the term a specific meaning, referring to the complementarity between the agent and the environment (as noted by Oliveira & Rodrigues, 2014).

humans and machines. When appropriately regulated, this relationship enhances certain human abilities, such as the capacity for natural language naming, and enables modes of existence that are only possible through the complementarity between agent and environment. This perspective positions digital technology not merely as a tool but as an integral part of the relationship that generates certain capacities in the human agent.

In the educational context, it is important to highlight the use of technology that complements the teacher's skills and enhances students' learning capabilities. Such resources are already distributed across the Web, through platforms that enable adaptive and personalized teaching, simulation environments using augmented reality, games, and virtual tutors. Recognizing the potential of these technologies for education involves acknowledging that we inhabit a new world, one that demands a new mindset aligned with the performativities (practices) emerging from people's interactions with these environments. This concept has been applied to the field of literacies by authors such as Jones and Hafner (2021), Barton and Lee (2015), and Cope and Kalantzis (2017).

Starting from the concept of *affordance*, Barton and Lee (2015, p. 45) observe that people create and are created by their environment. Thus, *affordances* are socially constructed, as long as people are able to perceive them and form ideas about how to act. According to Jones and Hafner (2021, p. 5), "humans are continually creating and adapting cultural tools to meet the needs of new material or social circumstances, or new psychological needs."

They argue that every cultural tool has *affordances* and constraints that, while enabling new actions, also restrict others. Based on this idea, they define *affordances* as "perceivable features in a cultural tool that facilitate the performance of certain types of actions," despite any constraints that may be created (Jones & Hafner, 2021, p. 192).

For example, a digital educational platform allows students to work at their own pace, regardless of space and asynchronously, but it restricts verbal conversation between them, limiting the sensory immersion that face-to-face interactions provide. The authors categorize the possibilities and constraints of the new environment into five types:

(a) What we can do;
(b) What we can signify;
(c) How we can relate to others;
(d) How or what we can think; and, finally,
(e) Who we can be.

Applying these topics quickly to the theme of the book, I suggest a few ideas:

- *First*: Digital educational platforms allow for flexible and simultaneous learning activities, complemented by AI, without the need for spatial and temporal synchronization.
- *Second*: These platforms enable multimodal communication, integrating writing, images, and videos, expanding the use and interoperability of natural languages;
- *Third*: Cyber-social environments offer new forms of connection and collaboration, forming online communities that transcend territorial boundaries and promote the co-creation of knowledge;
- *Fourth*: Digital technology expands human cognition, allowing for collective memory management and encouraging complex connections between knowledge;
- *Fifth*: Finally, these technologies can transform the social identities of teachers and students, promoting a more collaborative and participatory environment, redefining roles and dynamics in education.

Cope and Kalantzis (2017) explore the concept of affordances in a way that is closely aligned with the psychology of perception developed by Gibson, focusing on its potential for education. They propose seven affordances that have the potential to complement the capabilities of teachers and students in the learning process.

These affordances are defined in contrast to the constraints of conventional education, such as spatial limitations (the four walls of the classroom), temporal restrictions (fixed schedules), and participation limitations in discourse (only one student can speak at a time), among others. By recognizing these affordances, people can visualize new possibilities for action in their environment. As this action occurs, new affordances may emerge, as they result from the possibilities created by human creativity.

I present the seven affordances here, not because it is an illustration of the theory but because they manifest in the relationships that students developed during the courses I analyzed.

1. *Ubiquitous Learning*: This refers to the possibility of organizing formal education beyond the constraints of the four walls of a classroom and fixed schedules. In other words, learning resources can be available to students at any time and in any place, as long as connectivity conditions are ensured.

This possibility alone creates an unprecedented situation for education. It dissociates the learning relationship from the need for face-to-face and synchronous contact with a teacher in a confined space. This is highly disruptive. If taken seriously, it raises questions about the spaces and times that are still recognized and certified today as authentic spaces and times for learning, as well as questions the learning ritual centered on the synchronous speech of a teacher. Thus, this single affordance already opens a window of revision and possibilities to reimagine education.

2. *Active Knowledge Creation*: This suggests that students can use the creative resources available in software to generate knowledge. This possibility supports the pedagogy of design, which encourages students to appropriate existing knowledge and recreate it based on their interests and life circumstances. The process is facilitated by the abundance and diversity of learning resources available on the Web. Students can build artifacts throughout a course, with their learning process documented, paving the way for the practice of portfolios. The primary focus is on the knowledge artifacts they can create through writing, with secondary attention given to the underlying cognitive process.

3. *Multimodal Meaning Representations*: These go beyond conventional teaching methods based on passive reading and alphabetic writing, moving into the realm of multimodal digital media. Multimodality has always existed in languages, but this affordance highlights that it can now be recorded through computers, thereby expanding our understanding of text and writing. Once recorded, multimodal representations enter the world of designs, enduring over time and being reappropriated in various contexts. Working within this framework, however, requires a reevaluation of our understandings of language, linguistics, and learning. Initially, this potential demands an understanding of how mechanical language enhances the naming capacity of natural language.

4. *Recursive Feedback*: This suggests that all evaluation should primarily be formative, with summative assessment serving only as a retrospective look at the learning journey. Recursion can be utilized as a source of learning, encouraging different perceptual positions and opportunities for students to regulate their performance toward the course objectives. This possibility greatly expands the availability of feedback for learning, but it depends on changing the response times of conventional approaches. For feedback to be useful to the student's learning, it must be received

simultaneously with the activity. This approach can be facilitated by the use of big data, learning analytics, AI, and collaborative intelligence; in this context, knowledge is clearly seen as a social construct, and learning takes place in the social Web.

5. *Metacognition*: It suggests working simultaneously on curricular content, knowledge creation, and self-reflection. Technology enables the organization of learning processes that allow students to reflect within this broader scope. Here, recursion allows for the integration of multiple perspectives into the learning process, leading students to retrospectively analyze the outcomes of their actions in comparison to those of other students. When conducted at appropriate times, this comparison leads to self-reflection. CGScholar, for example, has a peer review design that facilitates working on metacognition.

6. *Collaborative Intelligence*: enables the reach and use of knowledge. This affordance demonstrates that once knowledge is stored on the Web, it becomes readily accessible, along with the people who can be engaged to contribute their expertise. The potential for collaboration is unprecedented. A classroom can operate from this perspective, encouraging students to develop skills that will be valuable in other areas of society.

7. *Differentiated Learning*: refers to a unique and social learning experience that leverages the diversity of students as raw material for learning. Differentiating learning according to each student's history, interests, and dispositions is something that many pedagogical approaches advocate. With technology, this potential can be achieved on a larger scale. In fact, each student can create their own learning path, and with the available technological resources, this journey can be supported and documented for analysis and record-keeping purposes.

I reiterate, however, that to work in an environment with such potential, an open mind is required—meaning a predisposition to the modes of existence that digital technology enables for people, as well as an openness to the expanded capacity of knowledge processes.

CHAPTER 3

The Methodological Path
of This Investigation

To develop my doctoral study and investigate the potential of the cyber-social environment in education, as defined in the previous chapter by the seven affordances proposed by Cope and Kalantzis, I worked with a specific platform—CGScholar. This platform was developed by researchers Dr. Bill Cope and Dr. Mary Kalantzis of the College of Education at the University of Illinois Urbana-Champaign (UIUC), based on the Multiliteracies Pedagogy perspective, and it has been applied in Brazilian universities in teacher education courses through a partnership between the American institution and the Department of Modern Languages at FFLCH-USP.[1]

More specifically, I investigated a multiliteracy practice that emerged from the interaction of these teacher-students through the platform. I believe that the results obtained from the application of CGScholar exemplify a complementary relationship between people and computers, and I hope these examples will inspire other educators to create meaningful practices with their students.

Experiences That Led Me to This Study

Before beginning this study, I worked for seven years on educational technology projects within the school environment. During this time, I gained experience with the main platforms available in the educational field, such as Moodle, Canvas, Blackboard, Edmodo, Google Classroom, and Apple Classroom. Additionally, I participated in an experimental project on blended learning, previously mentioned, with K–12 teachers from different regions of the country. This project received financial support from the Lemann Foundation and aimed to present practices that could inspire other teachers. At the time, I was part of a team of

---

[1.] Faculdade de Filosofia, Letras e Ciências Humanas da Universidade de São Paulo.

teacher trainers for the Edmodo platform and was one of the administrators of the digital community used by the group throughout the process. Subsequently, I took on the role of digital literacies teacher at a leading school in technological innovation in São Paulo, where I also contributed to teacher training.

Throughout this journey, I realized that we were immersed in instrumental approaches, akin to phonics-based literacy or functional literacy in the typographic context. I observed that our methods resulted in limited progress in pedagogical practices, yet I did not see alternatives beyond the strategies I was employing. Following this concern, I began to notice that first-generation learning management platforms reproduced a transmissive pedagogy that did not meet the demands for individuals who were collaborative, active knowledge creators, with analytical, creative, and critical skills.

This seemed to me an important problem to solve, rather than the need for mastery of hardware and software, as I had initially thought. This was observed later, through the study presented in this book, where I identify that the main challenge was and continues to be working with social learning relationships, classroom discourse, and practices of reading, writing, and assessment.

Redefining Education: The Role of CGScholar in Active and Collaborative Learning

I arrived at the university to investigate these challenges in 2017, and the initial path of the study was guided by the following question: how can a digital platform be created based on educational research to generate a learning experience aligned with the cyber-social perspective?

I believe that when we apply research in real educational contexts, we can observe needs and opportunities that can, in turn, be translated into the development of technologies that materialize them. On the other hand, I did not want to reproduce conventional pedagogies in the design of the technology. However, such an undertaking is beyond the scope of a doctoral research project. Therefore, I chose to investigate CGScholar, an educational platform that was already operating within this proposal.

The primary goal was to address my core research question. To achieve this, I studied the literature on the platform, as well as the field experience of its developers, educators, and researchers who use it in real teaching and learning contexts, according to the cyber-social proposal presented by Tzirides et al. (2023), which involves combining educational research with technology development.

My interest in studying CGScholar increased when I began participating in courses from the *Learning Design and Leadership* program at the University of Illinois, in partnership with USP (Universidade de São Paulo). At the end of a session of the New Learning course in November 2020, a student named Josefina da Silva published a testimonial on the course bulletin board reflecting on her learning process. I immediately recognized and identified with this testimonial. It was something I could have written, as it resonated with me after my initial participation in the courses:

> *I was impressed by the power of peer interaction that CGScholar provides, as well as the ability to learn through others' perspectives, which is truly enlightening. What is particularly constructive about the types of interactions this social media platform facilitates is the sense of empowerment that comes from exposure to different viewpoints and perspectives on a given topic. For instance, in the New Learning course, I was able to grasp and internalize various pedagogies and nuances implemented in different educational contexts—from police education in the US to elementary math education in Singapore, medical education in Bahrain, and special needs classrooms in the UK, among others. I was able to combine this with my previous knowledge and experience as an IT teacher in the UAE, and in doing so, I exponentially increased my confidence and ability to apply skills and theories in the classroom. In this case, the moment of classroom discourse where the teacher evaluates transforms into a collaborative assessment, as we all evaluate while engaging in peer reviews and commenting on updates. Being in a position where you are considered 'sufficiently competent' to give and receive relevant feedback is a huge motivational factor as a student and enhances your confidence in applying your own skill set.*

Josefina da Silva's experience with the CGScholar platform highlights the significant impact of **peer interaction** and **collaborative learning**. She is impressed by the ability to learn through others' perspectives, which she finds enlightening. The interaction provided by this educational social media not only expands Josefina's understanding of different pedagogies and global educational contexts but also empowers her by exposing her to various viewpoints. This testimonial reveals a learning process characterized by reciprocity and interdependence with her peers. At no point does Josefina appear as a student who stands out solely for her individual achievements.

During the New Learning course, Josefina explored teaching methodologies applied in various contexts and felt more confident and capable of applying theories and skills in the classroom. The platform transformed the traditional teacher evaluation moment into a **collaborative assessment process**, where students engaged in peer review and commentary, fostering a collaborative learning environment. This mutual feedback dynamic not only enhanced her skills but also reinforced her self-confidence in the practical application of her knowledge.

Although the student refers to competencies in two parts of the text, she does not state that the goal of the course was to develop them. Another aspect I want to highlight in this testimonial is her **meta-learning discourse**; she is reflecting on her experience in the course, trying to uncover which pedagogical resources led to the emergence and recognition of positive aspects in her professional identity. I consider this discourse to be an effect of metacognitive practice in peer review, which aims to prompt students to reflect on their thinking and learning. In other words, the student becomes aware of the subjective effects the learning process had on her, enhancing her understanding and practice of education.

I highlighted Josefina's testimonial to indicate that her learning experience reflects precisely what CGScholar proposes. However, I understand that it is important to provide some background on the history, technical aspects, and interfaces of CGScholar. Therefore, I will now focus on presenting and describing its functionalities. I have discovered throughout this work that many people are interested in learning more about this technology, what it enables, how it can be used, and how it might serve as a model for the development of other platforms. With this, I hope readers will gain a better understanding of the analyses I will present both in this chapter and in Chapter 4.

The CGScholar platform is rooted in a significant historical context for digital technology in general, and digital technology for education in particular. Indeed, the University of Illinois Urbana-Champaign was the birthplace of Mosaic, considered the first web browser, developed in 1992 by the National Center for Supercomputer Applications, a research institute based at the university, and PLATO (Programmed Logic for Automatic Teaching Operations), created in 1959, which was the first computer designed to integrate human vernacular in a digital media interface for formal educational applications (Figure 3.1).

The prehistory of CGScholar dates back to the year 2000, when professors Bill Cope and Mary Kalantzis decided to start developing software for writing and publishing on the web at education colleges in Australia, their country of origin. They already believed at the time in the importance of envisioning new possibilities for writing, publishing, and academic communication in the

Figure 3.1: PLATO prototype (workstation) from 1961.

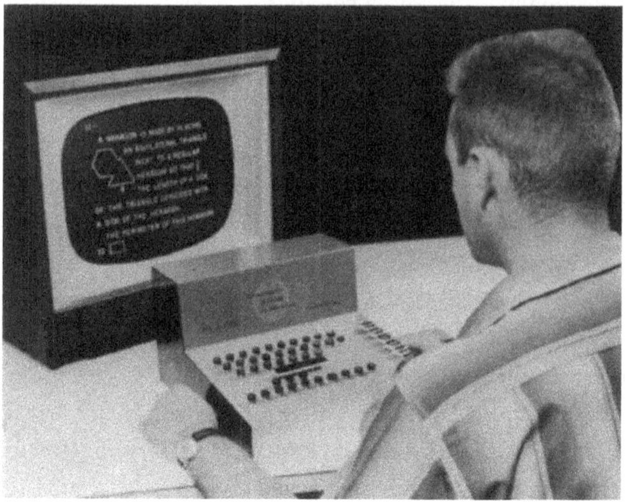

Source: Provided by researcher Bill Cope from the University of Illinois Urbana-Champaign. Available at: https://www.researchgate.net/profile/Bill-Cope/publication/351400910/figure/fig2/AS:1021002007990272@1620437256994/PLATO-I-workstation-1961.png.

context of digital media.[2] However, after a government transition in 2004, and with no access to resources, the programs were discontinued, along with other initiatives dealing with diversity in the country and integrating two research fronts—technological and multicultural.

In this context, they decided to act independently: "If people want to keep theory, research, and publications alive, they will have to do so on their own, outside of the system," argued Cope (Cope et al., 2005, p. 195). This decision laid the foundation for the educational research and software development project they have been conducting ever since.

When they began teaching at UIUC in the US in 2009, they initiated the creation of CGScholar. They also relocated the publishing and conference infrastructure they had previously worked with, Common Ground, to the UIUC Research

[2] They developed the CGPublisher project (related to academic communication in a digital environment) and ClassPublisher (focused on rethinking classroom discourse dynamics).

Park, renaming it Common Ground Research Networks and restructuring it as a non-profit organization.

In the first phase of CGScholar's implementation, experiments were conducted in the writing curriculum starting from the fifth grade.[3] During this period, existing computer-mediated assessment technologies were examined, and the potentials of big data and artificial intelligence (AI) were explored to enhance the assessment process. This study was applied to the development of the CGScholar Analytics app, which collects data on student activities and provides incremental feedback to learners, both through machine and machine-mediated interactions.

On the platform, fundamental concepts of the Pedagogy of Multiliteracies were incorporated into the design of its interfaces. These concepts had already been experimented with in classroom-oriented activity sequences during another project, Learning by Design.[4] Therefore, during the creation of the platform, there was a theory and a practice that were then redesigned in a completely digital environment. The development process of CGScholar follows the cyber-social perspective, combining Agile methodology[5] with design research in education, as cited in Chapter 1.

CGScholar is available for use, and until 2017, the platform's servers were located at Common Ground's headquarters in the UIUC Research Park. However, since 2017, the platform has been hosted on Amazon Web Services (AWS), and although the AI services offered by this provider were not chosen, AI continued to be developed internally.

In 2023, CGScholar hosted 350,000 accounts, with over 45,000 active monthly users. Its uses encompass literacy in elementary and secondary schools, higher education, including courses in education, engineering, medicine, and veterinary medicine, as well as global social learning interventions in partnership with the Red Cross and the World Health Organization (WHO).

The development of the platform has primarily been sustained through funding from foundations and government agencies, totaling $10 million since 2009. The business model adopted is "freemium," with most of the platform being open access and the availability of advanced tools for a fee. The aim of this model is

---

[3.] The reader can find references to this phase of the Scholar platform's implementation in an article published in 2013 by Cope and Kalantzis, titled "Towards a New Learning: The *Scholar* Social Knowledge Workspace, in Theory and Practice." Available at: https://journals.sagepub.com/doi/10.2304/elea.2013.10.4.332.

[4.] This project took place over the first decade of the 21st century and aimed to develop a Pedagogy of Multiliteracies based on classroom-oriented practices.

[5.] A methodology that became notable in software development and pertains to the processes involved in project management, encompassing people and tools that effect changes in the course of this dynamic.

to support the project sustainably, without seeking profit. This revenue covers the maintenance costs of the platform but does not cover the necessary investments for new projects and implementations in CGScholar, which continue to rely on funding.

Common Ground Research Networks finances the hosting of CGScholar's servers on AWS through revenues generated from conference registrations, journal subscriptions, and book sales. CGScholar also offers complementary applications, including "Event" (for hybrid conferences), "Bookstore" (for the publication and distribution of journals and books), and "Publisher" (for managing the peer review process).

From a pedagogical perspective, the design of CGScholar suggests changes in the architecture of learning, as it proposes a new classroom discourse that encourages active learning and collaborative intelligence as essential elements to rebalance agency between teachers and students. This new communication system, facilitated by the platform, can be understood under the concept of reflective pedagogy, in which students are viewed as active learners who engage through the interaction between their preexisting understandings, derived from their life experiences, and new knowledge that they themselves construct and reconstruct during the learning process. This dynamic is profoundly social, requiring constant dialogue among learners (as illustrated in Figure 3.2).

In CGScholar's learning framework, the focus is on the knowledge artifacts that students produce—their texts, which are expanded and multimodal— serving as manifestations of their thinking and evidence of their learning process. These writings can be of any textual genre. In my study, I worked with an academic essay proposal; however, a teacher can choose any other genre. One thing is the textual genre, and another is the dynamic used in its construction process.

What the platform promotes is a collaborative dynamic and access to perceptive and reflective positions that are not easily available in conventional writing processes. This includes the metacognitive position, in which students reflect on the design principles that are guiding their reflective activity while simultaneously practicing that same activity. Today, this position is considered essential for developing the forms of thinking necessary to understand the Web meaningfully, a dynamic that can be adjusted according to the teachers' objectives. There is no normative standard to follow. In fact, the platform is not exclusive to conventional classroom discourse.

In any case, the proposal for working with CGScholar is to articulate the availability of the platform's technological infrastructure with a theoretical

Figure 3.2: The Homepage of CGScholar, Modeled
on a Social Media Feed

Source: Screenshot, 2023. http://cgscholar.com.
Access to the Platform

framework created from affordances for cyber-social learning. In this sense, it is important to emphasize that in CGScholar, educational objectives are aligned with pedagogical principles and have been incorporated since the conception of its technical structure. In the following, I present some examples of various applications of the platform, before introducing the case studies in Chapter 4.

Some Features of CGScholar

When a user accesses the CGScholar platform, their starting point is an application called "Community," whose organizing principle is based on an activity feed, resembling social networks (as illustrated in Figure 3.2). This application provides a summary of the most recent activities conducted among participants and the various learning communities of which the user is a part. In CGScholar, users are not treated as "friends" and do not have "followers," as the platform considers that such concepts do not adequately apply to a learning community environment. Instead, they establish peer relationships. Horizontal relationships among participants are encouraged, and while they may resemble interactions in social networks, the platform suggests a distinct nuance for the educational context.

From the homepage, it is possible to access the community of a course (as shown in Figure 3.3). In this space, instead of the traditional distinction between teachers and students, users encounter the presence of administrators and community members. The term "admin" has become widely used to designate group coordinators on social media platforms on the Web. This approach also allows a student to assume the role of administrator or for a teacher to participate as an active member of the community, which is a way to work toward rebalancing agency between teachers and students in the sociotechnical dimension of the environment.

The alternatives for writing spaces in CGScholar encourage the production, sharing, discussion, and evaluation of public opinions through textual means. The unit of activity in a learning community is the update. Administrators create updates, which can be developed in real time or planned in advance. An update can contain text typed in the editor, attached PDF documents, embedded videos, audio files, or any other type of digital media. The software functionalities are similar to a blog post.

Community members are encouraged to participate in discussions in the comment sections related to the updates, following the dynamics described in the following.

Dynamics

Unlike the traditional classroom interaction model, in this environment, everyone has the opportunity to express themselves, either simultaneously or at different times. Participation in discussions can also be an essential course

Figure 3.3: Space for Posting Updates in the CGScholar Community

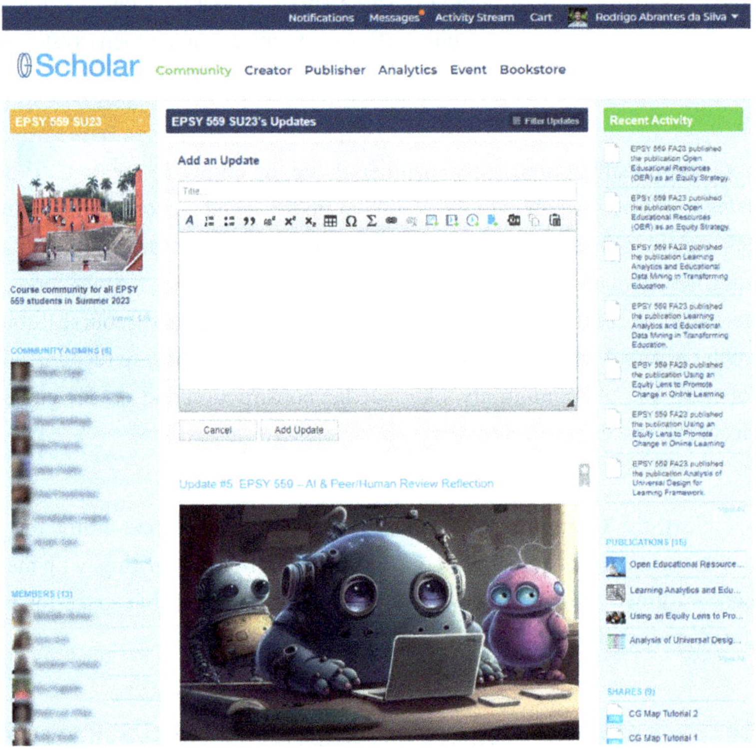

Caption: Interface for reading and writing in the course community. In this space, participants create and publish content related to the course topics. These publications are indexed in chronological order. By hovering the mouse cursor over the title of any publication and clicking on it, a new page opens with the complete publication, followed by a space for discussion through comments. To the right of each title, there is a marker with a star icon. By clicking on it, the publication is indexed in a user's favorites list for later access. To the right of the screen, under "Recent Activity," the latest actions taken by participants in the environment are listed. This space also allows the insertion of documents and links in the "Shares" block, located to the right of the screen and below "Recent Activity."

Source: Fragment from the wall in the Community environment.
Partial screenshot of CGScholar. Community: EPSY 559, 2023.

requirement, monitored and evaluated through the Analytics app, and even counted toward the student's grade and assessment. This approach is technically unfeasible in the chat areas of conventional learning management systems or during videoconference sessions.

Additionally, community members have the ability to create their own updates, contributing content related to their interests and personal perspectives, which are naturally diverse. The position of commenting on peers' updates and creating their own updates, with the classroom as a real audience, are two important channels for working on rebalancing agency in the classroom discourse through a structured and documentable dynamic. In this way, students are recognized as collaborators in the community's knowledge construction and co-designers of the learning process.

Next, the **Creator** app[6] It replaces the conventional text editor for extended assignments (Figure 3.4). It enables fully multimodal writing, with media embedded directly. Being cloud-based, it is a readily accessible space for peer feedback and teacher feedback.

The Creator screen is designed with a left/right division, along with a "curtain" that can be pulled from side to side to expand or minimize one side (Figure 3.5).

Activity Dynamics

On the right are peer reviews, self-assessment, instructor evaluation, rubrics, annotation tools—spaces to give and receive both human and automated feedback on the work. These feedback and interaction structures were designed to foster peer learning and formative assessment processes.

[6.] Creator is a space for text creation and peer review within the Scholar platform. The peer review system allows for the creation of rubrics (The term "meaning-making" refers to the process by which people interpret and make sense of their experiences, events, and information in their lives. This process involves the interaction between personal experiences, prior knowledge, cultural values, and social relationships.), which can be displayed on the same screen where the text is being written, on the right. This design option integrates the text creation criteria into the actual writing process, encouraging this dual layer of reflection from the production stage to the revision phase. Creator also connects to the Publisher app, which manages the peer review process through to the publication stage, at which point the work can be distributed on the student-authors' profiles, within course communities, and in the platform's bookstore.

Figure 3.4: The Creator app in CGScholar: Multimodal Representations of Knowledge

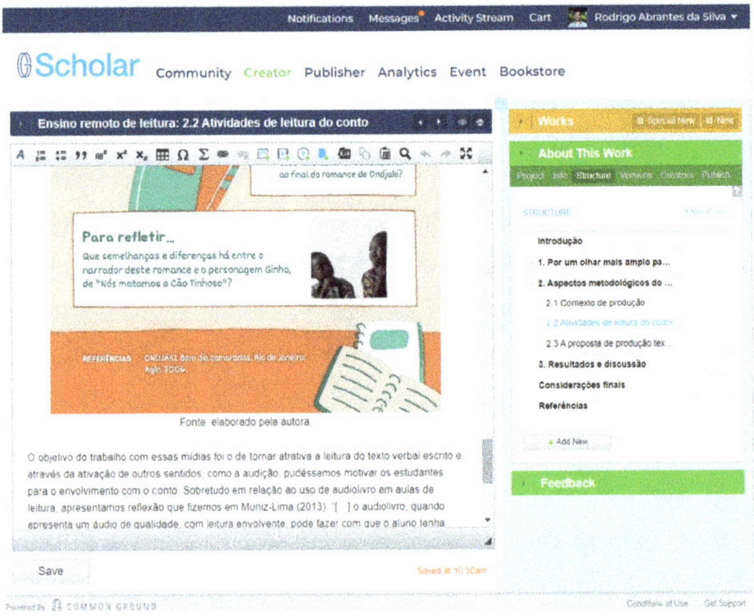

Caption: Screenshot of a stage in the creation of the work "Remote Teaching of Reading" by Maria José, with the section "Story Reading Activities" on the left side of the screen and the editing tool for the structure of the sections highlighted on the right. This work was created and published in the experimental journal *Digital Content Design for Education (DeCoDE)*, edited by the author of this study and Dr. Jailine Maiara Farias.
Source: Screenshot, CGScholar, 2023

Working in the Creator is a recursive process in which, throughout the various phases of a publishing project, students alternate between the left and right sides of the screen according to a logic that can be considered dialogic: between the work on the left and the feedback on the right; between learning on the left and assessment on the right; between individual thinking on the left and collaborative intelligence on the right; between domain knowledge on the left and disciplinary practice on the right; between content on the left and epistemological reflection on the right; and between cognition on the left and metacognition on the right (Cope & Kalantzis, 2023c).

Figure 3.5: Peer Review in CGScholar

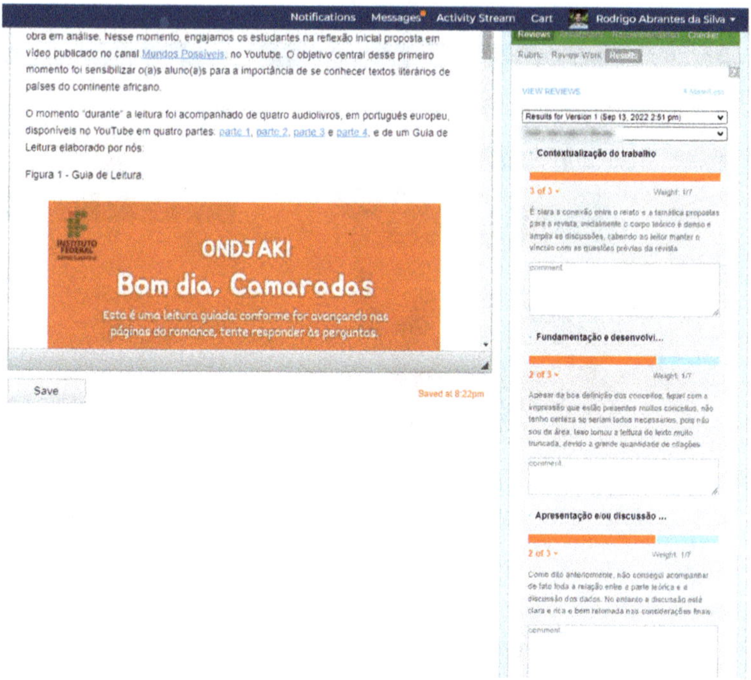

Caption: Screenshot of the work "Remote Reading Teaching" by Maria José, highlighting, on the right, one of the reviews of the author's first version, conducted based on a rubric developed by the journal's editors. For each criterion of the rubric, the reviewers assigned a point on the Likert scale and provided comments aimed at helping the author improve the writing of the work.

Source: screenshot, 2023.

[7.] We used an evaluation rubric on the platform. In my study, I applied seven rubric criteria for analyzing the works: (1) contextualization of the work, (2) foundation and conceptual development, (3) presentation and/or discussion of the object of study, (4) implications and applications, (5) use of media, (6) communication/organization, and (7) justification. This rubric was developed by me and Professors Dr. Jailine Maiara Farias and Dr. Vânia Castro to organize the publication process of two volumes of the journal DeCoDE. <Revista DeCoDE—Design de Conteúdo Digital para a Educação (cgscholar.com)>.

Subsequently, I used the same rubric to guide the writing and peer review process for students in the course "Linguistic Theories in Educational Contexts," taught by Professor Dr. Souzana Mizan. Each criterion in the rubric has a definition followed by a Likert scale—a widely used tool in social and psychological studies for measuring attitudes, opinions, and perceptions regarding specific topics—ranging from 0 to 3. This scale contains statements for the student-authors-reviewers to express their level of agreement with each criterion. In addition to indicating their level of agreement, there is a text box for each criterion where the student-authors-reviewers must provide a comment justifying the assigned score and offering guidance to help the student-author-reviewer improve their work. The complete rubric can be accessed via this link: https://docs.google.com/document/d/1D-Wwlgz662tMrOyJslQio92UDOWJNznARkIkVlQ7o4E/edit?usp=sharing.

This way of reflecting on the writing process aligns with the need for a meta-level understanding of the design being crafted—that is, thinking about the principles guiding the construction of the text. In this sense, Cope et al. (2005) argue that reflecting on the principles embedded in the design is an essential skill in the world of structural and semantic web markup, as well as the algorithmic operations embedded in the platforms available on the Web.

Researcher Gee (2014) complements this perspective, noting that a player improves their performance in games when they understand the principles of their design. This represents the capacity for reflection necessary to participate and be a practitioner in this new world of digitized text. To achieve this under-standing, digital literacies practiced in formal education should lead students to comprehend the design principles of languages.

Working on a project in CGScholar is a highly social and interactive process (Figures 3.6–3.8).

Dynamic Represented by the Flowchart

Main Framework: Stages of the writing and peer review process used in the Learning Design and Leadership program courses;

Figure 3.6: Collaborative Workflow and Incremental Formative Assessment in CGScholar, with Human Review and AI Review

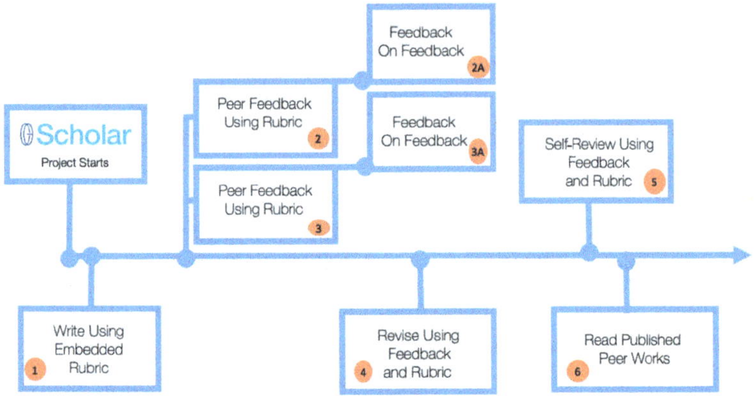

Source: New Learning Online. Available at:
https://newlearningonline.com/cgscholar.

Tables 2 and 3: Student-authors receive their peers' work for review, using the rubric as a guide to provide formative feedback;

Table 4: Student-authors revise their own work based on the feedback received from peers. They also continue to use the rubric as a reference.

Tables 2A and 3A: Student-authors reflect on the feedback received and evaluate it, providing comments to their reviewers. This stage allows student-authors to assess how useful the feedback was, and they can record their insights in the form of comments to the student-reviewers, who may use them to improve their review work. Simultaneously, student-authors reflect on how they created their own peer feedback, conducting a self-assessment of this process;

Table 5: Student-authors revise their own work using the same criteria applied by the student-reviewers;

Table 6: Student-author-reviewers can read the works published in the course community. Between steps 5 and 6, teaching assistants perform an additional layer of review, making adjustments to formatting and citations, as well as ensuring that student-authors have met the requirements for publication. After this review, the teaching assistants forward the works to the course instructor, who conducts a final analysis and decides whether they will be published.

In summary, before students even enter the writing project workflow, they begin interacting through updates and comments within the course **community**. Many discussions that start in the **community** space are further explored in the writing process of the project. Later, within the writing workflow in the Creator, they begin drafting on the left side of the screen with a rubric provided on the right, which guides the development of their texts (Figure 3.5). After receiving peer feedback, participants evaluate these reviews and revise their narratives accordingly (Figure 3.7).

Figure 3.7: Peer Reviews in CGMap: The Work Under Review on the
Left, Peer Review Map Based on the Rubric on the Right

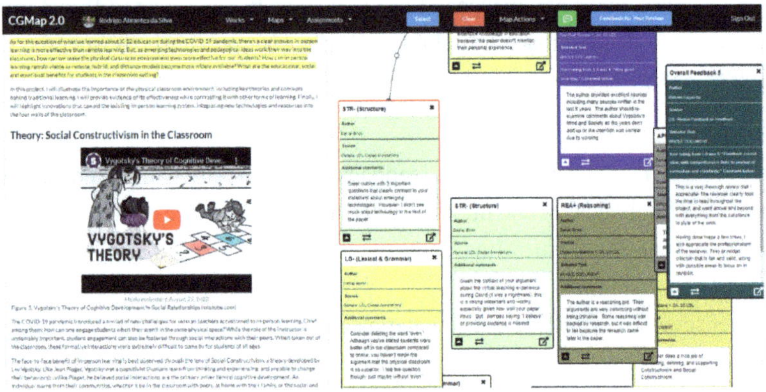

Source: CGMap. Printscreen, 2023.

At this point in the review assessment process, in January 2023, a new stage
was added to the workflow: AI review through CGMap (Figure 3.8).

Figure 3.8: AI Review in CGMap

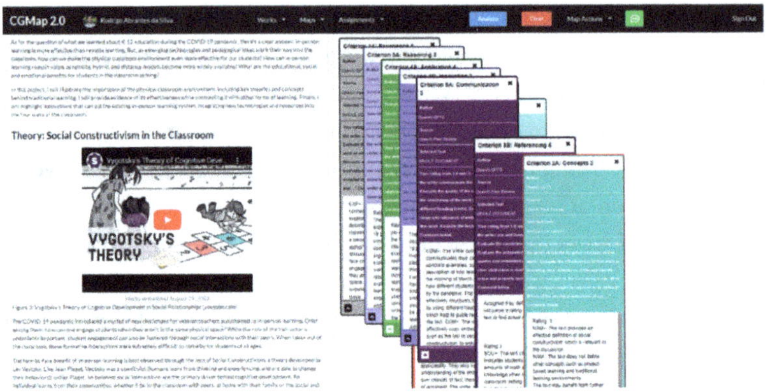

Source: CGMap. Printscreen, 2023.

AI Review Dynamics

Connected via API to OpenAI's GPT (Generative Pre-trained Transformer), CGMap provides feedback based on the same review criteria as human peers. Each of these stages requires a different perspective on the creation process.

Students compare human feedback with machine-generated feedback before the final revision and publication of their works in the course community and in their personal portfolios.[8]

After several months of experimentation and analysis of CGMap with students, this solution was refined. A database of 35 million words was created, comprising texts produced in the courses of the Learning Design and Leadership (LDL) program, as stored in CGScholar. These texts constitute a valuable repository of theoretical and empirical literature on pedagogy, literacies, and technological innovations in teaching and learning.

Now, with the language model of the generative AI of ChatGPT, a way has been found to utilize this archive, which has become a raw material for AI reviews. The proposed academic writing in the digital environment, as depicted in this book, thus becomes a learning resource, a tool for collaboration, and a means of accessing others' perspectives. It serves as a means to develop the cognitive and metacognitive skills necessary for community members to participate as practitioners of digital culture in an exploratory activity that fosters new argumentative capabilities, combining images, videos, and alphabetic writing.

By the time students reach the publication stage of their work, this rich experience of social learning has already occurred, such that the published work and the documentation of its literacy process retain the memory of the interactions, choices, and pathways that the student's thought has taken to arrive at the final outcome.

The Analytics application in CGScholar tracks students' learning through 22 different indicators (Figure 3.9). This is an automated feedback mechanism that facilitates the regulation and gradual improvement of the learning process. In such an educational environment, an AI tool can enhance human feedback by creating visual representations that illustrate how students understand the course objectives.

[8.] Cope and Kalantzis (2023c), Tzirides et al. (2023).

Figure 3.9: Analytics Application in CGScholar

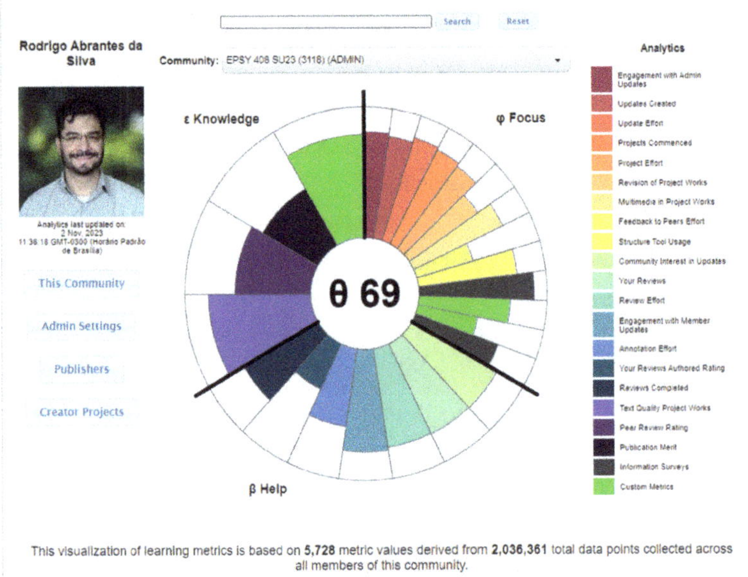

This visualization of learning metrics is based on **5,728** metric values derived from **2,036,361** total data points collected across all members of this community.

Source: CGScholar. Printscreen, 2023.

Learning Feedback Dynamics

The system analyzes and organizes data from students' interactions on the platform to create a type of visualization called the Aster Plot. Each stem of this visualization represents a learning indicator—something that the instructor has defined as necessary for the student to accomplish in the course (these indicators can be configured using a simple set of widgets). The indicators are categorized into three groups: "Knowledge," "Focus," and "Help."

In Figure 3.9, the aggregated data for all participants is available to the instructor. Additionally, each student has an individual visualization of their progress on their personal page, and the instructor can access these visualizations in the course administration environment. This is a way to work with automated feedback while learning is taking place. In a new course or set of learning activities, the student starts from zero and progressively advances toward 100, aligning with the instructor's learning objectives. There are no individual data points that the instructor can see that the student cannot also access, which aligns with the aim of a transparent and formative assessment.

Figure 3.9 consolidates over 5,000 actionable units of human and machine feedback, within a context where Analytics processed more than 2 million data points—far more than would be possible without computer mediation.

The result is a transparent and easily accessible assessment system that leverages big data tools and AI. Conventional learning management systems, with their hub-and-spoke file transfer architectures, lack the capabilities for such granular formative assessment or learning analysis.

Finally, in CGScholar, the Bookstore application serves as a repository for learning designs documented in a digital artifact known as a "learning module" (Figure 3.10).

Figure 3.10: Designing Learning in CGScholar: The Learning Module

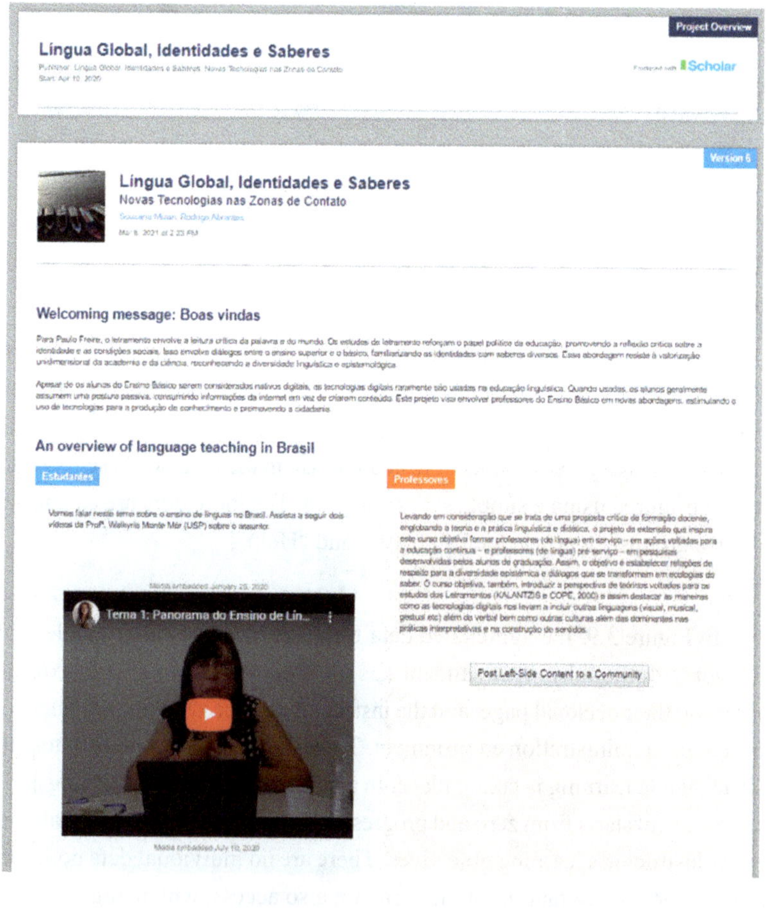

Source: CGScholar. Printscreen, 2023.

Module Dynamics

In the "learning module," instructors can create hybrid learning objects that combine traditional textbooks, syllabi, and lesson plans. The artifact features a two-column architecture. On the right side of the screen, a window facilitates communication among different educators active on the platform, allowing them to exchange information about learning objectives and processes. It also provides action buttons to share portions of content within course communities.

The analysis of the CGScholar platform demonstrates its pioneering role in transforming digital education by promoting an environment where collaboration and student agency are central. By integrating big data and AI, CGScholar provides an innovative and effective approach to feedback and formative assessment. The platform not only facilitates the production and sharing of knowledge but also redefines the dynamics between teachers and students, creating a more balanced and interactive educational framework.

My study highlights that CGScholar can serve as a model for future technological innovations in education, providing a robust alternative to traditional learning management systems. In the next chapter, the practical application of these functionalities will be explored through case studies, illustrating the platform's potential in diverse educational contexts. However, before we proceed, it is important to situate the context of my doctoral study by identifying the communities (courses) and peers with whom I interacted, a process that provided the elements for my final analysis.

On the Material Conditions of the Study

Pilot Partnership Between USP and UIUC

All my experience with CGScholar occurred within the scope of a partnership between the University of São Paulo (USP) and the University of Illinois Urbana-Champaign (UIUC), in two different contexts: (1) at the Cyber-Social Research Laboratory, where I had the opportunity to observe, experiment with, and analyze various courses from the Learning Design and Leadership Graduate Program at UIUC's College of Education, from which I extracted the analysis presented in the previous section; and (2) in a collaborative action research aligned with the objectives of the Common Ground Research Networks and the pilot partnership between the two universities, within the National Project on Literacies, developed in Brazil since 2009.

The national project, titled "Language, Culture, Education, and Technology," is registered in the directory of research groups of CNPq and involves collaboration among educational institutions from different regions of the country and foreign institutions.

Its main objective is to investigate the various contexts in which language teaching and literacy practices occur, aiming to develop local programs that promote learning from a critical perspective. My study represented one of the initiatives of the project, the results of which were periodically shared and discussed among the participating members.

The USP-UIUC Pilot Partnership is the result of a social design methodology (Gutiérrez & Jurow, 2016), focused on creating and expanding cyber-social learning environments co-constructed by international stakeholders. Gutiérrez and Jurrow argue that, in addition to seeking innovative approaches to improve education, this research methodology strives to make the design process part of social transformations and aligns with principles of social justice. For this reason, the National Literacy Project and the Common Ground Research Networks operate with a certain degree of independence from the university and governmental systems.

In my case, the integration into the National Literacy Project and the experience with CGScholar initially stemmed from my position as a tutor in the extension course "Multiliteracies and Teacher Education in Languages." The participation in the project allowed me to develop reflections on how schools and universities can integrate the pluralizations currently driven by digital epistemologies, based on the paradigms of digital and cyber-social literacies.

The Multiple Contributions of an Educational Platform Design

The integration between researchers through CGScholar provided me with the opportunity to conduct collaborative action research with Professor Dr. Suzanna Mizan from the Federal University of São Paulo (UNIFESP). In this type of research, the group participants work together to identify problems, develop solutions, and implement practical changes in a transdisciplinary process. Suzanna was participating in the "Multiliteracies" course in search of a platform capable of generating greater engagement among her students.

Throughout our collaboration, we converged on research focused on educational design, whose experience helped us: (a) build or deepen theoretical understanding; (b) shape the design of a solution for a real problem; and (c) propose multiple iterations of investigation, development, testing, and refinement, which can also be schematically classified into the following phases: analysis/exploration; design/construction; evaluation/reflection; and theoretical understanding.

For the purposes of the analytical scope of this study, I decided to follow two design cycles developed on CGScholar for two regular courses taught by Prof. Suzanna, one undergraduate and one postgraduate, offered by the Department of Languages at UNIFESP Guarulhos, in São Paulo area.

In this type of action research, the position of the researcher is active, going beyond observation and data collection. In my case, it was nurtured by the guidelines of ethnographic studies, which are based on the principle that "the researcher always has some degree of interaction with the situation being studied, affecting it and being affected by it" (André, 2012, p. 30). It is a process marked by continuous co-creation between the subject and the object of knowledge, where the emphasis is on "what is happening rather than on the final product or results," and the goal is to discover new concepts and relationships.

The partnership with Suzanna was characterized by constant dialogue, in which we shared views, bibliographies, and action perspectives, which converged in the practice of the courses. In Chapter 4, I reconstruct parts of this experience through narrative elements, built from the effects that our dialogue had on me and my retrospective analysis of what was recorded from that experience (including Suzanna's pedagogical actions with her students during the regular courses at the institution).

CHAPTER 4

Cyber-Social Learning and Multiliteracies: Case Studies

Meaning is an act of social cognition.
Bill Cope and Mary Kalantzis (2020)[1]

The extension course "Multiliteracies and Language Teacher Education," introduced at the end of the previous chapter, was based on the CGScholar platform and video services like Zoom. It represented the opportunity I found to study the contours of the cyber-social learning environment proposed here, providing the group—tutor and teachers in training—with access to updated readings on literacy theories, focusing on the discussion of problems and the development of classroom-oriented activities.

Intervention in a Local Context

The course was structured around updates on English language teaching, including multiliteracies, critical literacies, multimodality, and the integration of digital technologies in education. In each update, various sources were distributed, such as videos recorded by the course instructors, links to external websites, images, and academic articles. The verbal texts of the updates articulated these diverse materials, adopting a hybrid approach that incorporated elements of conventional writing and orality to bring it closer to the spoken discourse in the classroom.

---

[1] Cope and Kalantzis (2020).

Participants were encouraged to expand on the themes of the updates through three discursive stances:

- In the comment space at the end of the course instructors' updates;
- In creating their own updates, connecting their experiences to a course theme and thereby producing content representative of their perspective, which would become a type of text available for learning in classroom discourse; and, finally,
- Through encouragement for participants to read and comment on their peers' updates.

The objective of this dynamic was to create a structure of collaboration, a support network, and a distribution of feedback among participants. Additionally, students had opportunities to practice publishing texts and other expressive resources in a web community, considering what and how to publish based on a group of readers. I believe this is a way to regulate the act of publishing on the web, giving it a reflective dimension.

In each of these positions, the course stimulated the active involvement of participants in knowledge construction and valued each individual's experience as raw material for learning from a dual perspective: (1) as the participant's learning, who reflects on their experience in relation to the course theme, and (2) as learning for a peer, who learns by reading the reflection produced by a colleague.

This discursive coordination corresponds with literacy practices observable in digital media environments and with the epistemological practice presented in Chapter 1 of this book, which consists of locating epistemic authority in personal experience and individual voice, which are foundational to the culture of participation. In other words, the logic of participatory discourse observable in digital culture has replaced the standard classroom discourse, transforming the conventional, hierarchical model centered on the authority of the teacher into a cyber-social learning discourse that is heterarchical[2] and distributed to each participant in the learning environment.

[2.] Unlike a hierarchy, where there is a clear structure of command and control, heterarchy operates with greater flexibility and dynamism, allowing different elements or members to assume leadership at different times. This can promote a more egalitarian collaboration and a rapid adaptation to new circumstances or information. Heterarchies are seen as useful in environments that are complex and constantly changing, where a more rigid and hierarchical approach may be less efficient. For example, in some modern organizations, project teams may function in a heterarchical manner to solve complex problems, leveraging the specific skills and knowledge of each member.

The proposal aligns with the perception of the researchers cited here that conventional pedagogical practices no longer engage students as they once did, making it difficult for teachers to connect with the new generation. It is essential to acknowledge the contributions of participants in their written productions. Thus, the course was not aimed solely at offering ready-made knowledge but rather at valuing participants' contributions to their own development and production. This is an essential practice to align subjective dispositions with digital culture because, for this environment to thrive, it needs to be recognized in people's subjectivity.

In this cyber-social learning environment, I maintained constant dialogues with Professor Suzanna. Right in the first update from the instructors participating in this training course, she shared the following observation: "I believe that my students [at the University] who learned English on their own using new media are much more successful than the effects that new media have on the educational system."

In her observation, she first emphasizes that, in her area of expertise (teaching and research in English Language), students are demonstrating better performance outside the classroom, through digital media, than during conventional classes. Conversely, Suzanna argues that new media has not had the same positive impact on the educational system, which continues to reproduce conventional pedagogical practices. Even with the incorporation of new media, the educational system still adopts epistemologies developed for a society that did not have computers or the internet. Thus, she highlights the central problem of the lack of correlation between teaching and learning practices in educational institutions and social practices in other contexts of everyday life, such as in the workplace or in community life.

This problem has been debated for three decades by scholars linked to the sociocultural paradigm of literacy, such as those already mentioned: Cope and Kalantzis (2000), Lankshear and Knobel (2011), Gee (2014), and Monte Mor (2017). Although the COVID-19 (coronavirus disease 2019) pandemic forced the school community to use digital technologies, the problem persists, indicating that it is a result of structural processes shaped by long-standing historical experiences, giving them such rigidity that they are still perceived today as something "natural" in society. I refer again to the discourse of the conventional classroom, which finds its maximum expression in the centrality of the lecture format, along with all the institutional apparatuses that compose and sustain it.

This inherited structure includes deeply rooted conceptions of what constitutes a class; how learning occurs; the class hours; the time dedicated to lessons; the moments and spaces designated for reading, writing, and discussions; the course topics; the adopted bibliography; the evaluation methods; publications; and teachers' compensation, among other aspects that shape the education system. These elements derive from discourses created in specific historical processes and cultural contexts. Therefore, their reproduction is not a natural act but rather a result of the agreement and active participation of the individuals involved in the educational system.

Suzanna expands her reflection by asking how it is possible to construct knowledge in the classroom based on the freedom each student finds to follow their own path toward the "knowledge they are interested in building." These two conceptions—"constructing knowledge in the classroom" and allowing each student to "follow their own path"[3]—represent fundamental pillars in contemporary understandings of education. However, the big question lies in how to implement a pedagogy that effectively enables this perspective, as these ideas often remain mere statements of intention when they reach the classroom.

Suzanna believes that digital technology offers a pathway to work within a framework that encompasses such objectives. She acknowledges, however, that many bureaucratic processes hinder the advancement of this approach, precisely because they reproduce practices and concepts structured for another time and another society.

Throughout the course, I was struck by Suzanna's stance on the platform: she took every opportunity to express her position, both in writing her updates and in commenting on her peers' updates, as well as in her engagement in discussions. Moreover, she used many multimodal elements in her writing, incorporating various argumentative functions that expanded verbal writing. In her first update, she presented a collection of images produced by her university students for an assignment that involved capturing images of establishments with names in English and analyzing them from a critical perspective, as seen in Figure 4.1.

Residents of the São Paulo Metropolitan Region, Suzanna's students documented, through their images, as she herself explains, the spread of the English language as an invasion, a kind of "territorial marking." From a critical standpoint, her

---

[3.] "Following your own path": This proposal does not mean adopting an individualistic stance. On the contrary, the interdependent relationship with the learning community is fundamental to the learning experience.

Figure 4.1: Update from Professor Suzanna, Showing an Image Created by a Student

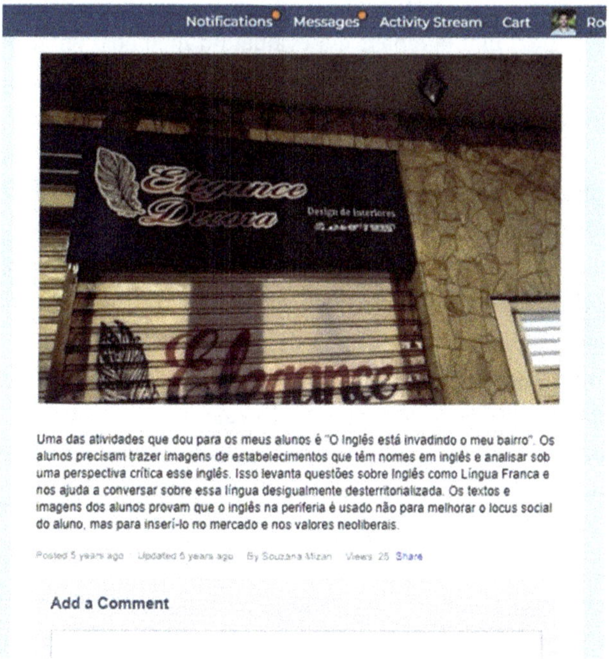

Caption: "One of the activities I assign to my students is 'English is invading my neighborhood.' Students need to bring images of establishments with names in English and critically analyze this use of English. This raises questions about English as a Lingua Franca and helps us discuss this unequally deterritorialized language. The students' texts and images show that English in the periphery is not used to improve the student's social position but to insert them into the market and neoliberal values."

Source: CGScholar. Screenshot, 2023.

students concluded that this marking aims to recruit individuals into the market, a process that occurs through identification with the values projected by this language, as mentioned by Suzanna in her conversation with another course instructor (Figure 4.2).

In this context, Suzanna's own students, who also work as English teachers in basic education schools or language schools, play a crucial role in the process of

Figure 4.2: Example of Comments in the Discussion Section of the Update Shown in the Previous Figure

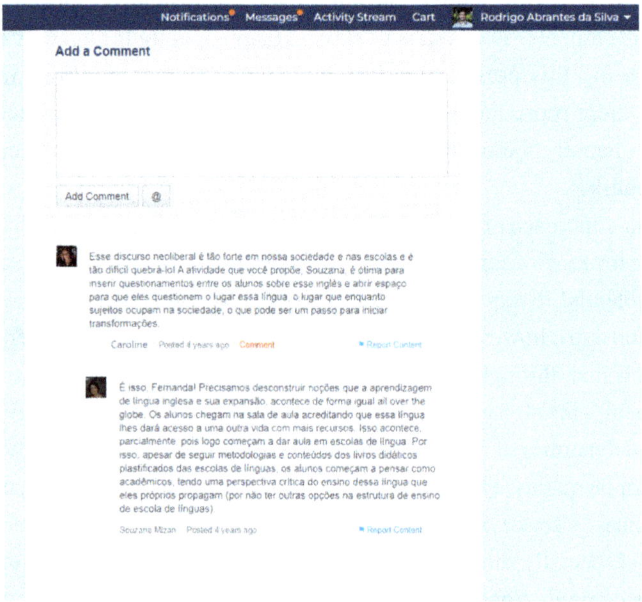

Caption: Comment 1: "This neoliberal discourse is so strong in our society and schools, and it's so hard to break it! The activity you propose, Suzanna, is great for prompting students to question this use of English and opening space for them to reflect on the place of this language and the place they occupy as individuals in society, which can be a step toward initiating transformations."

Comment by Souzana Mizan [Suzanna Mizan]: "That's it, Fernanda! We need to deconstruct the notion that the learning and spread of English happens equally all over the globe. Students come into the classroom believing that this language will give them access to a different, more resourceful life. This happens, in part, because they soon start teaching in language schools. Therefore, despite following methodologies and content from the polished textbooks of these schools, students begin to think like academics, developing a critical perspective on the teaching of this language that they themselves are spreading (due to the lack of other options within the structure of language schools)."

Source: screenshot, 2023.

"recruitment" and conversion of individuals.[4] They promote the assimilation of the values disseminated by the instructional materials of these institutions in their classes. From this perspective, Suzanna's activity fosters a critical awareness among the students. The goal is to expand their capacity for agency, enabling them to bring this perspective into their own classrooms, breaking away from the mere reproduction of the role that conventional discourse assigns to them—namely, being mere agents of value transmission or recruitment into the market.

Suzanna's approach of seeking new purposes for teaching the English language resembles my relationship with digital technology. Just like language, technology has the potential to expand meanings and broaden access to new ways of life and relationships; however, in the current context, without a critical perspective, the result is that, through technology, we lead people to reproduce in the digital environment—through platforms and services of large market corporations—the values and dynamics of the predominant economic system in society (with all its relations of inequality, exclusion, meritocracy, low wages, informality, etc.). Just like Suzanna, I also see myself working within the very discourse that technology projects—especially through the big techs of Silicon Valley—trying to modify it from its contradictions and possibilities.

The challenges related to education in the digital environment were experienced on a large scale and with significant impact during the COVID-19 pandemic. It has been widely recognized as one of the greatest crises faced by educational systems throughout history, according to António Guterres, secretary-general of the United Nations (UNESCO, 2022). The disruption of in-person activities exacerbated existing inequalities and highlighted well-known historical problems, such as the lack of access to digital infrastructure and the inadequacy of conventional pedagogical approaches for online teaching (UNESCO, 2022). It was a historical context that tested the ability of institutions, teachers, and students to adapt.

Based on the experience gained from my study, it is important to highlight that, from the beginning, Suzanna and her students did not face difficulties with the digital environment that could not be resolved through conversations or messages to use the proposed platform. Thus, it is likely that the pandemic favored the technological appropriation process by her students for using the necessary

---

[4.] The conversion of individuals was a foundational practice in Brazilian society. As a colony of exploitation, the first to systematically adopt this strategy were the Jesuits. They were also responsible for creating a language that served as the basis of communication in the colony and for establishing an educational system. These initiatives lasted until their expulsion in 1759, during the government of the Marquis of Pombal (Malerba, 2020).

tools, as digital became the only available resource for those wishing to remain isolated to avoid the risk of contracting the virus.

For the students, at that moment, the platform represented a tool through which they could achieve the course objectives and progress in university. However, I observe that this outcome was only "partially" generated by the pandemic, as Suzanna, for example, had already been determined to find a digital approach for her projects and had never faced difficulties in using the CGScholar environment.

The context of the pandemic established a circumstance that, from the socio-cultural perspective of literacy (Gee, 2014; Street, 1984/2014), made the social experience of this period conducive to the development of digital literacies. Just as, from the perspective of this study, reading and writing do not occur without specific purposes, the use of technology, like the use of language, also does not. People perceive and appropriate resources to meet social purposes, not intrinsic purposes of the resources themselves. Social purposes motivate people to discover resources and mobilize skills to achieve them. In the next section, I will describe and analyze other courses offered by Suzanna to her university students, using platforms such as CGScholar and Creator.

New Horizons: Pedagogical Practices in the Cyber-Social Environment

Between November 2020 and February 2021, Suzanna began teaching the course "The Decolonial Option in Language Teacher Education" to a class of 47 undergraduate students. She guided the students to focus their participation through posting updates on the course community's wall. I prepared a tutorial to instruct them on how to create an account on the platform and joined one of the classes at the beginning of the course to help Suzanna teach them how to use the text editor for publishing their updates.

Suzanna decided not to engage with all student comments in the discussion sections of the updates, which is a common practice in the facilitation of distance education forums.[5] She made this decision to give more weight to peer comments

---

[5.] In Brazil, it is common for tutor teachers to be instructed to mediate all comments in distance education forums. However, recent studies indicate that student learning is more effective when the teacher does not constantly perform this mediation. For this approach to work, the course must be designed in a way that encourages active and collaborative student participation in these environments. Therefore, the teacher's absence in direct mediation does not signify a reduction in teaching work but rather a shift to a more complex level, focusing on course design, analysis of student work, and personalized guidance.

on updates, recognizing that this approach would be a way to enhance students' agency in classroom discourse and promote horizontal communication from a heterarchical perspective. This group produced a rich body of written work.[6]

Subsequently, between March and August 2021, now in a graduate course for 30 students, Suzanna continued to build on this approach,[7] Suzanna decided to expand her approach on the platform by adding a peer review project. She named this project, which would serve as the final assignment for the course, "Decolonizing Language Teaching Journal." The goal was for students to produce academic articles. At this point, they assumed the roles of both writers and reviewers for one another.

The project lasted six weeks, was structured in Creator (Figure 4.3), and was divided into the following stages: draft, feedback, revision, and publication. Once each student-author submitted a draft of their first complete version in Creator, the work was sent to two peers, who were tasked with reviewing it based on seven rubric criteria.[8] After receiving feedback, the author would revise their work and submit it for publication. At this point, Suzanna would read all the students' articles and assign their grades.

I observe that by the time they reached this stage—submitting their work to the teacher—the participants had already engaged in dialogue about their productions, received feedback, and revised their work based on peer comments. They became a support and motivation network for each other, which sustained their writing process.

Methodological approaches of the described courses

The goal of **peer review**, as an experimental phase of the research, was to provide constructive and forward-looking feedback to enhance students' learning outcomes, represented by the multimodal writing artifacts they produced. The dynamics of this project involved several experimental approaches:

First Approach: Regarding the writing artifact, the proposal was the production of a multimodal article, meaning that students were required to

---

[6] This work is detailed in: Abrantes da Silva and Mizan (2022). Available at: https://cgscholar.com/bookstore/works/decolonial-practices-on-the-educational-platform-cgscholar.

[7] "Linguistic Theories in Educafictional Contexts."

[8] See note 7.

Figure 4.3: Work Construction in the Creator App

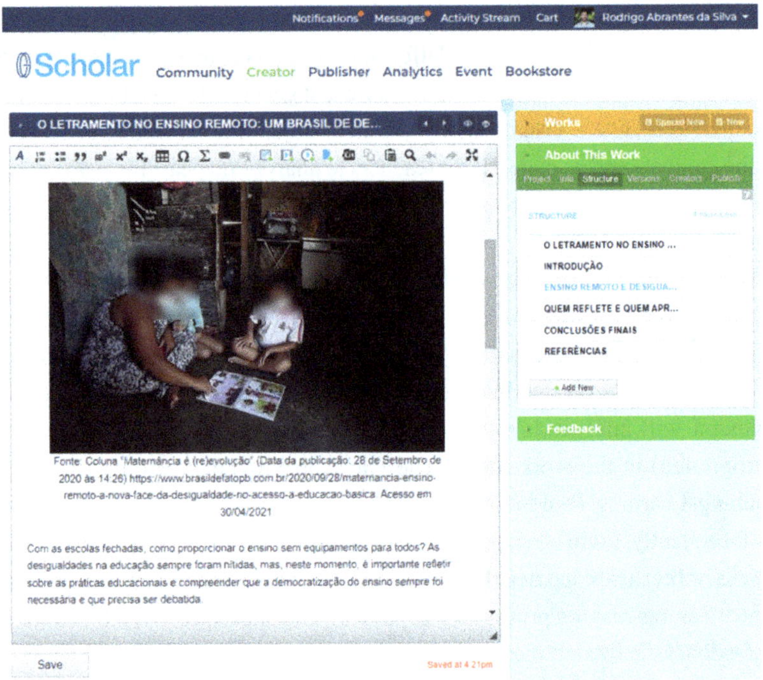

Caption: Screenshot showing the development of the project "Literacy in Remote Learning: A Brazil of Inequalities" by student Juliana Nogueira Lopes, a participant in the course "Linguistic Theories in Educational Contexts," taught by Professor Suzanna in the first semester of 2021, during the university closures. The students in this course completed a peer-reviewed writing project using the Creator app.

Source: CGScholar. Screenshot, 2023.

work with verbal writing alongside other (multimodal) forms of knowledge expression;

Second Approach: Writing was also used as a mode of interaction and collaboration among students during the peer review process, aiming to balance the receptive mode of reading with the productive mode of writing;

Third Approach: The data from the writing process and student interactions were recorded through the platform. Part of this data was processed by the Analytics learning analysis system, a feature that generated semantically readable visualizations. These visualizations served as an additional source of feedback for the students and as material for teachers to analyze the learning paths of each student and the group as a whole.

To provide the reader with context on how the possibilities for collaboration and reflection in writing were expanded through the opportunities enabled by the platform, I emphasize that these opportunities represent precisely the concept of affordances, introduced in Chapter 2.

Next, I will analyze how these affordances emerged (and were utilized, appropriated) in the work titled "The Hip Hop Movement: Rap Culture as a Teaching-Learning Tool in Schools," by student Anna Falchi.

Subsequently, I will analyze this excerpt from Anna's writing process through the lens of **literacies** and **decoloniality**. Initially, I highlight certain characteristics in the meaning-making process of the student, and subsequently, I illustrate how the feedback facilitated the expansion of some of these characteristics.

In the version that Anna submitted for review, she proposed the following introduction:

> *The concerns that gave rise to this writing emerged during the course "Linguistic Theories in Educational Contexts," taught by Professor Souzana, who uniquely led me to reflect on various topics such as: subject, culture, identity, coloniality, decoloniality, modernity, and other issues addressed and discussed in synchronous meetings that frequently emerged unexpectedly in my life.*
>
> *In this context, I approach Hip Hop culture as a tool for contributing to decolonial pedagogical practices. I present the rhymes and rhythms of Rap as a way to confront racism, violence, silencing, and the erasure of the peripheral subject, which are produced and reproduced within society.*
>
> *Finally, I emphasize Rap as a break from colonial academic concepts, capable of promoting autonomy, empowerment, and assisting in the decolonization of thought and knowledge.*

From the "concerns" raised by the themes discussed in the course, Anna develops a social identity as a "peripheral subject," incorporating elements of the hip-hop movement and rap. These forms of expression allow her to establish connections with her community, coinciding with the location of her university, and weave these elements into the context of the course, as she mentions: "I approach Hip Hop culture as a means to contribute to decolonial pedagogical practices."

This perspective exemplifies the observations of Monte Mor et al. (2021), who highlight the current generation of literacy in Brazil with an emphasis on the social perspective, enriched by studies that interlink history, culture, and society. The approach is also aligned with Monte Mor's (2015) reflections on critical formation, where she suggests the importance of learning that is not merely passive but encourages students to critically reflect on the world around them. Additionally, the initiative of raising awareness, de-naturalizing, and decolonizing social differences can be in harmony with ideas of social justice, as advocated by the author. This social focus is emphasized in the decolonial approach present in Suzanna's course, as cited by Anna.

The decolonial perspective allowed for the emergence of a critical discourse from the student regarding "colonial academic concepts." She identifies a "third space,"[9] an intersection zone shaped by coloniality, here represented by rap as a cultural expression of the peripheral subject, historically subalternized in dominant narratives. For Anna, rap is a vehicle for social justice, providing her with a repertoire to combat racism, violence, silencing, and the "erasures of the peripheral subject produced and reproduced within society."

In the educational context analyzed here, the "third space" emerged as representations of social identities in classroom discourse, transforming into accessible knowledge for the student community, who became the real audience of readers for the texts produced throughout the course.

Before proceeding with the analysis of Anna's work, it is important to conceptually highlight some aspects of writing in the contemporary digital environment.

Exploring the Boundaries of Writing in Web 3

In the sociocultural view of literacy, writing is considered an embodied social practice (Gee, 2014; Street, 1984). The word "embodied" is used when its use

---

9. The term was coined by sociologist Ray Oldenburg (1989) to describe places other than home (first space) or the workplace (second space), spaces where people go to socialize with others beyond their coworkers and family. In the field of Language and Education and Applied Linguistics, the most well-known use of this term comes from the studies of Bhabha (1998).

can occur unconsciously.[10] In this context, writing begins to be applied in various ways without people necessarily being conscious of these uses and their purposes. In other words, it becomes part of associative and reflective processes that are not always conscious. Gee (2020) expands this perspective by explaining that embodiment occurs when our thoughts and practices intertwine with those of others, as well as with our own experiences. We form associations, often without consciously understanding how they happen. This is where a broader form of intelligence begins to operate, often without our knowledge. This is a feature of the complexity of the human mind.

This complexity allows humans to significantly expand their capacity for reflection, as the unconscious part of the mind works without individuals being focused on it. Being aware that this unconscious functioning exists and being able to consider it in conscious processes, such as decision-making, for example, can help people solve problems, visualize possibilities for different life paths, and think creatively and critically. I understand that practices that promote self-reflection, such as language studies, philosophy, and psychoanalysis, are areas that develop the ability to handle the expanded mind.

Such practices have always been scarce in societies; however, in the current context, they are even more so, and I perceive that a symptom of this is the subordinate position of the humanities in the organization of basic education curricula. I also consider that the scarcity of these practices favors the design of commercial platforms, which function to capture and direct people's attention—in other words, they reduce human agency to transform people into objects of algorithms. To better explain this statement, I request permission here to expand the sociocultural perspective of literacies, bordering on recent readings of cybernetics through anthropology.

As Cesarino (2022) aptly demonstrates, commercial platforms were designed to operate on instances of the unconscious. From this perspective, these processes are linked to emotions such as love and hate and are activated by immediate forms of

---

[10.] Alerto o leitor que a palavra "inconsciente", em função da divulgação da psicanálise fora de círculos acadêmicos e/ou psicanalíticos, pode ser objeto de várias apropriações, muitas das quais o associam a algo "desconhecido" e/ou "irracional". Não é nesse sentido que a emprego. Na falta de palavra melhor, mantenho aqui seu uso para especificar uma dimensão outra em relação ao saber consciente, aquele sobre o qual a pessoa sabe, sendo também um saber, porém sobre o qual pouco se sabe. Em uma perspectiva que combina filosofia da mente com letramentos, Gee (2020) mantém o uso de *"unconscious"* para nomear processos com essas características. [I warn the reader that the word "unconscious," due to the dissemination of psychoanalysis outside of academic and/or psychoanalytic circles, can be subject to various appropriations, many of which associate it with something "unknown" and/or "irrational." I do not use it in this sense. For lack of a better term, I maintain its use here to specify a dimension other than conscious knowledge, that about which a person knows, which is also a knowledge, but about which little is known. From a perspective that combines philosophy of mind with literacies, Gee (2020) maintains the use of "unconscious" to name processes with these characteristics.]

communication, favored by the emotional appeal that the platforms' multimodal communication provides. Regarding this trend, my position as an educator is to provide access to reflective practices that develop human agency and distance people from algorithmic manipulations. However, for this "immunization" to occur, people need to develop this reflective practice in the Web environment, about which they must acquire discernment and awareness.

If the platformized Web delivers "to users personalized worlds that confirm their individual frameworks—in cybernetic terms, containing an excess of positive feedback," as Cesarino (2022, p. 105) states, the question that arises for a literacy practice that enables people to be agents and not simply objects of platform algorithms consists of developing aptitudes that displace them from the location where platforms expect to find them. For this, they need to reflect on an additional layer of language, which allows them to act consciously on the causal regime operating in Web spaces.

Reflections on Writing, Metacognition, and Agency

Metacognition refers to the ability to reflect on one's own thinking (Cope & Kalantzis, 2017) and promotes the development of agency (Wen et al., 2023). In the present study, we identified that the ability to reflect on thought was fostered by specific resources. First, the use of writing helps make thinking processes visible. Moreover, the criteria established in the rubric encourage cognitive effort. These same criteria allow the different works to be compared by their authors. They function as one of the elements that constitute a "baseline," that is, a starting point against which subsequent results can be compared.

In this regard, I observe that this comparison occurs in the minds of the student-author-reviewers when they are performing the revision work. Thus, it becomes possible to access the thought process in question and use interpretation as a form of meaning-making,[11] in the form of feedback, to support learning. It is not, therefore, a comparison aimed at establishing a ranking between works. Such comparative reflection, especially during the revision phase, allows authors and reviewers to understand each other's thought processes, thus contributing to the visualization of their own ways of thinking. In this way, unconscious processes are brought to awareness from a common baseline, contributing to

---

[11.] The term "meaning-making" refers to the process by which people interpret and make sense of their experiences, events, and information in their lives. This process involves the interaction between personal experiences, prior knowledge, cultural values, and social relationships.

cognitive expansion. This dynamic represents the metacognitive circuit explored in this book.

I return to Anna's work to elucidate these issues. In Figure 4.4, the student's text is displayed on the left, while on the right, a review is shown with emphasis

Figure 4.4: Work in Progress in the Creator App, with the Review Process Highlighted on the Right Side of the Screen[12]

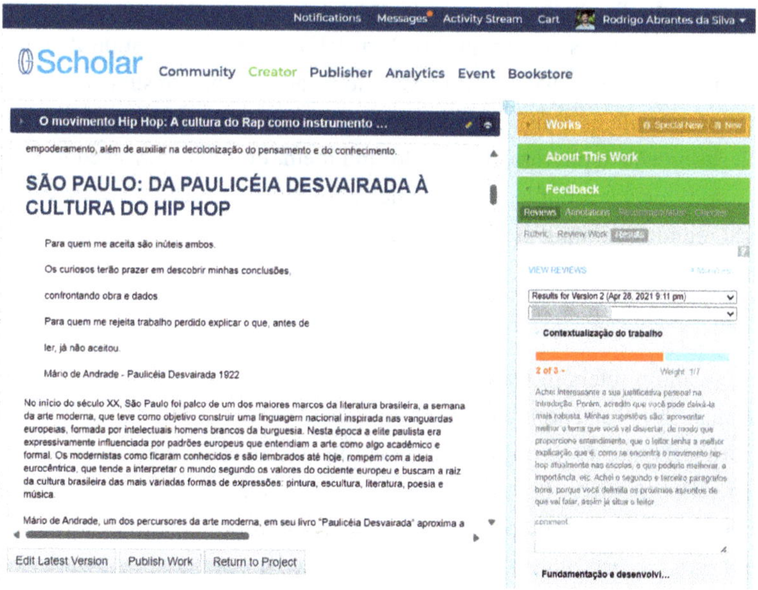

Caption: Student Anna's text on the left:

> For those who accept me, both are useless.
> The curious will take pleasure in discovering my conclusions, comparing the work and data.
> For those who reject me, it is wasted effort explaining what they have already refused to accept before reading.
> Mário de Andrade–Pauliceia Desvairada–1922

At the beginning of the 20th century, São Paulo was the stage for one of the greatest milestones in Brazilian literature, the Week of Modern Art, which aimed to construct a national language inspired by

---

[12] The group's name and images were excluded due to image rights concerns. The caption "the MCs" aims to situate the portrayed profile within the hip-hop scene.

European avant-gardes, formed by white bourgeois intellectual men. At that time, the São Paulo elite was heavily influenced by European standards, which viewed art as something academic and formal. The modernists, as they became known and are remembered today, broke with the Eurocentric idea that tends to interpret the world according to Western European values and sought the roots of Brazilian culture in various forms of expression: painting, sculpture, literature, poetry, and music, one of the forerunners of modern art, in his book."

Reviewer's rubric on the right, with emphasis on the comment she made on the criterion "Contextualization of the work":

"I found your personal justification in the Introduction interesting. However, I believe you can make it more robust. My suggestions are to better present the theme you will be discussing, so it provides a clear understanding, offering the reader the best explanation of what it is, the current state of the hip-hop movement in schools, what could be improved, its importance, etc. I found the second and third paragraphs good because you delineate the topics you will address next, thus already situating the reader."

Source: CGScholar. Screenshot, 2023.

on the reviewer's comment regarding the contextualization criterion of the work. The reviewer offers praise and constructive criticism, expressing initial interest in the text and subsequently proposing improvements. For example, she suggests a more in-depth introduction, clarifying the theme and the current state of the hip-hop movement in schools, proposes rewriting in some areas, and highlights the importance of others. Moreover, she praises the second and third paragraphs for effectively delimiting the topics to be discussed, thus guiding the reader:

> I found your personal justification in the introduction interesting. However, I believe you can make it more robust. My suggestions are: better present the theme you will be discussing so that it provides a clear understanding, giving the reader the best explanation of what it is, the current state of the hip-hop movement in schools, what could be improved, its importance, etc. I found the second and third paragraphs good because you delineate the topics you will address next, thus already situating the reader.

In this interaction, the reviewer interprets what her attention captures from Anna's text and collaborates to make certain aspects visible to the student-writer.

This is already an interweaving of thoughts, experiences, and practices, which materialize through writing and can serve purposes such as providing a basis for further elaborations by the students or the teacher, for analysis and feedback, or even for testimony, record, or evidence of learning.

I highlight that consideration for the reader is one of the elements that brings this writing closer to a literate practice, shaping a discursive situation different from the conventional model of text production in courses at various levels, where students write only for one person—the teacher. I also observe that the evocation of the reader here is not merely a rhetorical device for the student to "imagine" a reader but rather refers to the course participants as her potential readers (who also act as collaborators and, in a way, co-designers).

The reviewer's feedback motivated the student to keep writing, meaning that a human reviewer exercised agency and, in doing so, became a stimulus for the expansion of the student-writer's text. In the revised version, Anna added two more paragraphs based on the previous reviewer's comment:

> *Hip-hop culture was born in the United States in the 1970s when major cities faced a severe deindustrialization crisis, plunging the population into a violent urban environment: drug consumption, racism, police oppression, state injustice, and, above all, social inequality. The greatest victims of this scenario were the most economically vulnerable: Black and Latino communities. In Brazil, Hip-hop arrived with a transnational perspective, meaning its theoretical orientation comes from another country. Starting in the early 1980s, hip-hop culture initially emerged in Brazil by valuing Afro-descendant identity through black music parties.*
>
> *Comprised of rap, graffiti, and breakdance, hip-hop has moved from the margins to the classroom. Through this artistic representation, teachers develop learning in the field of education.* **Tapping into students' interests gives meaning to the study and captures everyone's attention, allowing for multiculturalism and social diversity in the educational institution, paving the way for interdisciplinary discussions in subjects such as History, Physical Education, Geography, Art, English, Portuguese, among others.** *(our emphasis)*

The student enriched her analysis with historical context information, establishing connections with social processes such as "a deindustrialization crisis" and "a violent urban environment." This addition allowed Anna to suggest an interdisciplinary perspective for her classroom work. In the second paragraph, she deepened her understanding of the relationship between hip-hop and the school environment, arguing that "it is through this artistic representation that teachers develop learning in the field of education." This statement shows the impact of the feedback received, which motivated Anna to improve her analysis of the interconnections between hip-hop, rap, and school education.

This experience stimulated the process of metacognition, both in the author and the reviewer. Both reflect on the nuances of writing in light of the design principles of this activity, that is, the criteria established by the rubric, along with the prerequisites of text production. This aspect also illustrates one of the points of significant collaboration in the course, highlighting the role of writing as a resource for learning and critical reflection. Language, in this case, was used to mobilize reflective thinking, linking experience and social identity.

Peer review through the CGScholar platform facilitated the application of classifications and comments based on an integrated rubric, as well as offering mechanisms for feedback on feedback (expansion leading to metacognition). I would like to remind you that we are analyzing teacher training courses for undergraduate and graduate students in this digital teaching environment, which they, in turn, will apply to their students, many of whom are public school teachers in basic education.

The experience of learning from the writing of other students provides the student-reviewers with the opportunity to analyze and compare their peers' writing with their own. This practice allows student-authors, regardless of their prior experience, to benefit from the review process, triggering a reflective analysis of their own work based on peer reading. It helps to understand the principles that govern the writing process. In this scenario, an intense sociocognitive effort takes place, fostering the formation of an epistemic community.

In order to understand how peer review, a practice that emerged and consolidated in the scientific publishing field, can be used as a pedagogical resource, I propose a historical perspective on this practice.

Peer Review in Education: Origins and Contemporary Perspectives

Peer review is essentially a feedback mechanism, fundamental for knowledge construction. It is a common practice in academic journals. An article submitted for publication is sent to different researchers on the subject, with an equal or higher level of expertise than the author. It is an extremely valued practice and is used as a quality standard in scientific publications. Various modes of review and feedback can be found among writing teachers, writers, and editors. Finding someone who can provide good feedback is valuable. At the same time, good feedback serves as a source of motivation to acquire the practice of revising texts and continuing to write.

Over the past few decades, several authors have studied different peer review models, and great emphasis has been placed on its pedagogical use, not only in writing courses but as a possibility for all courses where writing is used for some type of report, essay, or any other textual genre.

History

The practice of text review dates back to ancient times. Hooper (2019) investigated the publication process of Cicero's works and identified a figure, Titus Pomponius Atticus (110–32 BC), who played a role similar to what we would now associate with an editor. His duties included critiquing style and content, discussing the suitability of publication, and evaluating the appropriateness of titles. Atticus also conducted private readings of new books, sent complimentary copies, and organized their distribution. Due to the scope of his responsibilities, he can be considered one of the first editors in history.

In the context of Islamic science, Cope and Kalantzis identify textual practices that follow a course of action similar to modern peer review. For example, in the work *Ethics of a Physician*, they identified a procedure mentioned by the author Ishāp bin Ali Al Rahwl (854–931), akin to the type of evaluation sought in peer review: a physician's notes were examined by a council of doctors to determine if a patient's treatment adhered to appropriate standards.[13]

The institutionalization of peer review occurred mainly after the dissemination of the printing press. The journal *Philosophical Transactions of the Royal Society*, considered the first scientific publication in the West, formalized the process. Authors such as Moxham and Fyfe (2018), on one side, and Chapelle (2014), on the other, disagree on the exact start date of this type of review, establishing it respectively in 1655 and 1731.

In any case, it was only after World War II that peer review experienced significant growth, especially during the 1950s and 1960s, when scientific research funding increased. In this context, peer review became a policy of excellence and a quality seal for journals, serving as a screening process to identify, through

---

[13.] Meyers (2004), Spier (2002) cited by Cope and Kalantzis (2009, p. 32).

consensus among reviewers, the most suitable contributions for the editorial line (Tennan et al., 2017). The expansion of this practice is also related to the specialization of knowledge areas and the increased volume of articles that began to be evaluated by qualified experts.[14]

With the advent of digital media, the opportunities to manage the peer review process have increased significantly, primarily due to electronic resources for submitting and receiving texts, as well as for documenting reviews. However, digital media has also brought other possibilities for this practice. As argued by Tennant et al. (2017), it has facilitated its use as a resource for formal learning.

The use of peer review in the classroom allows for aligning pedagogy with some characteristics of cyber-social learning.[15] For example, when practiced by students instead of experts at the top of the career hierarchy, it aligns with the culture of participation (Shirky, 2014), making each person's contribution relevant to the course and transforming feedback into an abundant resource in the learning process. Secondly, it modifies assessment, shifting it from being exclusively centered on the teacher to being distributed among the students. This allows individuals to take responsibility for their own learning, the learning of others, and, consequently, for the course community. By facilitating a multiplicity of discourses, diversity emerges and becomes visible in classroom discourse. From this perspective, these new dynamics contribute to rebalancing agency[16] and establish principles of horizontality in learning relationships.

In conventional education models, evaluating students' work is a unidirectional action, flowing from authority (the teacher) to non-authority (the student). In these models, the student's position is not addressed in its agentive dimension, and the intentionality attributed to their actions does not aim to generate effects in the real world. However, by adopting peer review, a broad field of possibilities opens up, as when students take on the role of reviewers, they not only learn to conduct quality reviews but also to work based on the feedback received.

Benefits

Therefore, among the benefits of peer feedback, the literature[17] highlights that it brings a genuine sense of audience to the classroom, something that provides

---

[14.] Cope and Kalantzis (2009, p. 34).

[15.] Cope and Kalantzis (2022b, 2023b).

[16.] Cope and Kalantzis (2016, 2017).

[17.] Kaufaman & Schunn (2010), Montebello et al. (2018), Keh (1990), Pinheiro (2020).

purpose and a sense of authorship—unlike when the only readers and reviewers are the teachers themselves—and helps to develop students' critical reading and analytical skills. Critically reading peers' drafts leads to improvements in students' attitudes toward writing and learning. Moreover, it encourages them to discuss alternative viewpoints that may lead to the development of new ideas.

Researchers in the field further emphasize that peer review values cultural differences among students, proposing that these differences be seen as raw material for learning. It promotes more improvements in writing than teacher feedback and enhances reading and comprehension skills. Additionally, a body of studies indicates that access to peer feedback can help address learning issues stemming from socioeconomic background.[18]

As for its implementation, some authors[19] note that the review process must be carefully structured, and training on its execution is necessary to ensure success. During the review practice, students take on various roles, with the collaborative stance being the one that most favors meaningful revisions. The literature[20] highlights the importance of addressing students' expectations regarding peer work and the role of the teacher in the classroom. If students adopt defensive positions, resist collaboration, or disrupt one another, the quality of the work may be compromised. However, when properly guided, it becomes a formative, collaborative, and recursive process that can lead to reflection, responsibility, learning, change, and/or improvement.

More Recent Research

From the perspective of the pedagogical use of peer review, a study published in 2019[21] focused on analyzing the conditions that favor the transfer of skills during the review process, especially when based on rubric criteria. From the perspective adopted in this book, rubric criteria should be designed to represent the skills desired for student development. The researchers argue that the cognitive work performed during the analysis of the criteria, as well as during the comparison of the different applications of these criteria in peer review, contributes to the learning of the targeted skills.

[18.] Hattie and Timperley (2007).

[19.] McGroarty and Zhu (1997), Stanley (1992), Villamil de Guerrero (1996).

[20.] Carson and Nelson (1996); Francis (2021); Wen et al. (2023).

[21.] Joordens et al. (2019).

The research also revealed a comparative relationship between peer reviews and expert reviews, indicating that the average of multiple peer assessments aligns with expert evaluations. It is also relevant to highlight that the 2019 study showed that students who reviewed their peers' work developed a more accurate understanding of the quality of their own work. The authors conclude that this approach, applying rubrics to peers' written work, can be adopted in various learning contexts, promoting a shift in student perspective and providing meaningful learning opportunities.

This experience allows student-authors, regardless of their experience, to understand the principles guiding the writing process. In this context, intense sociocognitive work occurs, shaped by the contours of an epistemic community in which a reciprocity ethic is developed. Students begin to see themselves as co-responsible for their peers' work, fostering interdependence and interconnectedness in relationships.

Moreover, the process of reviewing multiple papers at each stage of a course can enhance students' understanding and interpretation of the rubric, offering an analytical and practical perspective. Students can employ their analytical skills to deepen their understanding of a topic and reflect on the different ways an academic paper can be structured. Finally, peer review, especially when conducted anonymously, involves critical interaction with the text. In the cases analyzed in this book, the review is an integral part of formative assessment:

> In teaching and learning, peer review has the advantage of training learners to participate in these canonical knowledge processes. In conventional educational terms, peer review provides a channel for formative assessment. It also contributes to the formation of epistemic community, where students learn to give and take on board constructive feedback. (Tzirides et al., 2023, p. 92)

Cope and Kalantzis (2016) propose the development of writing assessments aligned with "21st-century skills," such as active engagement, participatory citizenship, and innovative creativity. Pinheiro (2018, p. 11) notes, in turn, that this peer review model "makes it possible to work with concepts of design and agency...through more horizontal levels of interaction in formal educational settings."

With modern computational analysis approaches, including machine learning, data mining, and natural language processing, both writing and review data can be personalized and analyzed. Another point to consider is that the Digital Age allows for the management of the peer review process in ways previously impossible in the pre-digital format. With computational resources, it has become feasible to capture the complex epistemic performance of writing and visualize it through representations, as we will explore later (in the section addressing "Assessment in Literacies: Beyond Testing" and presenting examples in Figure 5.4). As a result, the use of peer review enables the dissemination of knowledge to a much broader audience than traditional teaching.

Social Literacies and Skill Development in Peer Review

As extensively documented in the literature, the pedagogical use of peer review is now a common practice in various contexts of formal education. Among the authors referenced here, I find it important to engage in a discussion with the study of Joordens et al. (2019), who advocate for the metacognitive work facilitated by peer review as a means of skill development. Their study focuses on the cognitive aspect of this process and is grounded in an individualistic view of cognition. This perspective, however, contrasts with literacy theories in several essential aspects.

Street (2014, p. 148), for instance, sees an irreconcilable dichotomy between "social practices of reading and writing" and the development of skills. For him, any attempt to approach literacy from notions of "competence" would be equivalent to abstracting it from social practice and transforming it into a "thing," thereby reifying literacy. Similarly, Lankshear and Knobel (2011) reinforce this argument by asserting that in order to develop literacy, one must be embedded in social practice. Based on these arguments, I understand that literacies can lead to skill development without, however, reducing them solely to a means of transferring such competencies. This would mean stripping literacies of their essential element: the intelligence embedded in social relationships.

In the context of my doctoral study, peer review was used pedagogically, interwoven with a sociability that took shape within the learning community. Students mobilized capabilities to use the platform, analyze, and formulate strategies for interpreting course themes, as well as to consider the possibilities of structuring an academic paper. In this process, student-writer-reviewers underwent the experience of assuming different identities to engage in collaborative intelligence work. Each of these social roles involved a distinct type of dialogue about text construction, promoting reflection from various perspectives and configuring a metacognitive activity that, indeed, led them to develop certain skills. However, these skills were not the ultimate goal of the process.

While the model proposed by Jordeens and Paré advocates for the "transfer" of skills, this book argues that students should engage in a journey aimed at having each of them publish an article by the end of the course. Moreover, while the cited authors emphasize individual cognition, I propose social cognition, rejecting any form of measurement of intelligence detached from its social context. In this perspective, I reiterate that literacy practices can lead to the development of competencies and skills, but this is not their primary goal. Competencies and skills are not developed through the mere assimilation of concepts about them. In a literacy practice, what matters is not the learning of concepts, but the specific activities that individuals engage in and how they relate to one another to achieve social purposes.

If we emphasize skill development, we risk transforming social relationships into something abstract. We must exercise caution to avoid this pitfall, especially because the peer review process can stimulate the development of competencies such as learning regulation, often associated with "student autonomy" in contemporary education. In the experience analyzed in my doctoral thesis, students developed skills to meet the writing prerequisites for the academic essay, involving new learning to use the digital platform, articulate course themes, enhance their writing, and provide constructive feedback during the learning process.

It is essential to highlight one final point regarding this development: when applying a pedagogical model in a structural and social context different from the one in which it was created, it is crucial to ensure that the process does not occur in a top-down manner, where teachers and students must adjust to external criteria and/or criteria that are alien to them. In this study, we sought to model the CGScholar platform to align with the course objectives and the students themselves. Therefore, education in the digital environment should always operate from this perspective, considering the social context in which a given experience occurs. Thus, the practice created in one context can be adapted to another without losing its essence, as long as the structural similarities between the contexts are maintained and the social and cultural variations in its application are respected.

Re-signifying Education: Social Mind and Decolonization

"The human mind does not learn by storing
generalizations and abstractions.
It learns via experience."
J. P. Gee (2014)[1]

Let us return to our case studies, reiterating that these reflect an experience in a digital environment, the CGScholar platform, which enables dialogue with diverse realities and forms of expression, reflecting on all of this from the social place, the projects, and the expectations of the community and each of the students.

In writing from the cultural movements mentioned, Anna's writing recreates her message within the educational environment, and both become means of breaking away from "colonial academic concepts." Indeed, by proposing this representation of the peripheral subject in the form of an academic essay, Anna integrates it into the universe of available designs and inscribes this memory within the dominant discourse, as I have previously noted. I want to emphasize that this integration reflects one of the main tenets of the multiliteracies theory: "that a theory of meaning as transformation or redesign is also the basis for a theory of learning."[2]

I notice a connection between the pedagogy of design and the results generated by the juxtaposition of discourses I found in Anna's text.

In this way, Anna realizes what Suzanna had expressed in one of her comments published in the course "(Multi)literacies and Teacher Education." Suzanna wrote: "In developing pedagogical practices that facilitate occupying a classroom as a space for the exchange of knowledge, digital media can play a role in creating virtual environments where a variety of knowledge can circulate."

---

[1] Gee (2014).

[2] Kalantzis et al. (2020, p. 176).

In this process, a community is formed that recognizes itself through literacy practices, within a set of linguistic transactions aimed at stimulating the learning of specific course topics, always in dialogue with the history and experiences of the students. The knowledge produced throughout this process becomes valuable materials that can serve as learning objects for other individuals and communities, as they can be shared on the Web after publication.

I believe this is the point where digital tools allow for the amplification of awareness processes in the classroom, especially through access to representation, resulting in the construction of identities, the possibility of inscribing one's story in the collective digital memory, and gaining recognition through the experience of otherness that arises from reading. I also think that each of these movements has been sought by critical pedagogy, as well as by the Pedagogy of Multiliteracies.

I find it essential to emphasize the processual and social aspect of this experience. I also perceive this proposal as closely aligned with the concept of learning as embodied cognition, as formulated by Gee, who asserts that "the human mind does not learn by storing generalizations and abstractions. It learns through experience." In my study, I observe that students, through the experience of revising another's text as per the suggested guidance, identified certain aspects that caught their attention, wrote about them (feedback), and "edited" their reading experience, thereby creating an interpretation. Or, in Gee's terms, they created a mental model that they can utilize when writing and/or revising their own work.

Comparing the writing project in question with Gee's (2014) theory of learning, the process represented a way for students to construct meanings through experience rather than merely through other words or definitions. This is how a skill outlined in a rubric criterion—specifically, the contextualization of the work in this case—can be interpreted by a student, which in no way implies that all students will arrive at the same interpretation. On the contrary, each will construct their own understanding, and the dialogue between student-authors and reviewers will revolve around the individual experiences of each person transformed into written representations.

This process aligns with the logical sequences in which learning occurs according to Gee's perspective: (1) first, students perform a task by writing; and (2) subsequently, they read with the objective of improving their own writing and assisting others. As Gee asserts, "real learning is doing before reading. Reading to do better. Doing comes before competence, not competence before doing."[3] In this conception, language is used to generate meaning about specific

---

[3.] Gee (2014, p. 91).

practices and problems. There are no meanings outside of this relationship and the practices to which the language refers.

Therefore, when students stated that reviewing each other's work helped them reflect on their own (Figure 5.1), this indicates that:

They had, while writing, the opportunity to compare the mental representation they created based on one of the rubric criteria with the representation created by the other.

Exposure to difference prompted reflection and expanded their understanding of the scope of the criterion itself, opening channels for new elaborations (This impact is demonstrated in Graph 1).

In these terms, the criterion for analyzing both the other's work and one's own functioned as an operator. That is, it enabled the development of a cognitive or sociocognitive model, fostering the regulation of learning, metacognition, and skill development through performance.

This is an example of the "social mind" in practice, where the digital media interface and computer-mediated human dialogue are constitutive parts of this mind: they are artifacts of embodied cognition.

Gee deepens his argument by highlighting connections similar to those I found in Anna's text. He states: "Saying (language) and doing (action) are also inseparably

Figure 5.1: Students' Perception of Peer Review Work

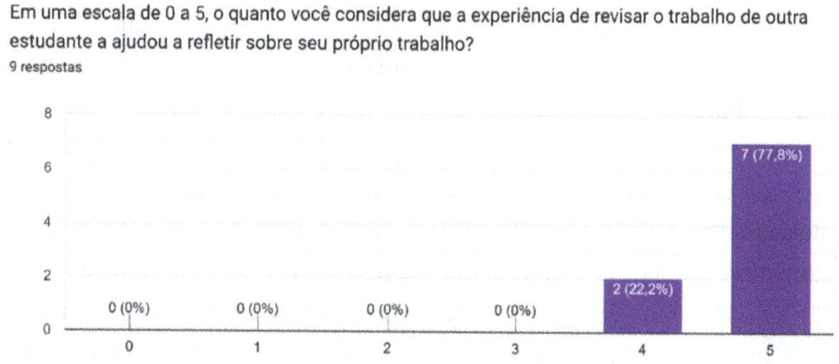

Em uma escala de 0 a 5, o quanto você considera que a experiência de revisar o trabalho de outra estudante a ajudou a refletir sobre seu próprio trabalho?
9 respostas

Source: Prepared by the author, 2023.

linked to being (identity). By 'being,' I mean assuming a socially recognizable and meaningful identity."[4] This is what I found right at the introduction of the student's text: an intrinsic connection between her saying, her doing, and her being, represented in this context by the identity of the peripheral subject. She presents herself as a teacher motivated to fight against the "erasures" that are "produced within society" through "racism, violence, and silencing."

In this writing process, identity does not emerge as something that belongs to the student as an individual. On the contrary, her writing reflects a clear social context: there are the rap artists, the students, her teacher, the university; in other words, following Gee, there are "people and institutions that create the conditions for the…recognition" of an identity that is profoundly social.[5]

I would like to focus a bit more on Anna's work analysis. From the perspective of another concept analyzed by Gee, namely, embodied cognition, I observe that the proposed writing project enabled the student to work from a representation available in her memory about the peripheral subject—not a representation derived from academic discourse. Her work consisted of editing this memory, formatting it into the creation of a new academic discourse, one that is non-colonial, as the student herself emphasized. The formatting of this new academic discourse occurred through a particular relationship with memory and operated a form of "cognitive justice." To better illustrate this operation, I refer to the work of Mignolo (2003).

This author argues that alphabetic writing has historically operated as a device of colonization, as it produced a silencing effect on the memory of Indigenous peoples and, consequently, on enslaved Black peoples.

> *Silenced societies are, of course, societies in which there is speech and writing, but they are not heard in the planetary production of knowledge, which is guided by the local histories and languages of the "silencing societies" (that is, the developed ones). (Mignolo, 2003, p. 108)*

Through the imposition of their language, the colonizer devalued local languages and cultures, imposed their values, and established epistemic hierarchies that sustained colonial power. From this perspective, I understand that the writing project I am analyzing presents a path of "epistemic decolonization," enabling a certain "reparative" effect in relation to the coloniality highlighted by Mignolo.

[4.] Gee (2014, p. 91).

[5.] Gee (2014, p. 93).

The effect of restoration consists of enabling silenced representations in memory, which have been obscured by dominant discourses, and writing them through dominant languages: alphabetic writing and academic discourse.

Inscription in Collective Memory: Expanding the Frontiers of Knowledge in the Digital Age

The final phase of the writing project analyzed in my study consists of the publication of the texts. Making them public is an essential step to complete the design cycle. It is the moment when the new design enters the world of available meanings, being able to be read and used by anyone who has access to it. This is also the stage where there is a transition from the limited context of the classroom to the broader instances of society and culture. It represents a direct connection between classroom work, characterized as a knowledge-building process, and the return of the result of this work to society in the form of knowledge as a common good, through an academic essay available on the Web. The Web serves, at this stage, as the environment that will host the new knowledge artifact (Figure 5.2).

This approach also provides an example of how classroom production can be documented in portfolios. For teachers, the set of records encompasses the entire writing process and the data from student interactions in the writing spaces of the platform, which they can collect and generate visualizations (Figure 5.2). In other words, the learning outcomes compose a portfolio of student production, which evolves from being merely a demonstration of what was done in the course to becoming knowledge available to the community. This practice provides a pathway for the contextualized production of knowledge, where students can apply the resources offered in the courses to issues relevant to their locality.

Figures 5.2 and 5.3 show excerpts from the work "The Revitalization of Indigenous Language Through the Lens of Decolonization" by student Carolina Dametto.

In this context, Carolina's work complements Anna's by expanding the space for enunciation and resistance against dominant colonial narratives and structures. Suzanna's course prompted the student to seek engagement with indigenous languages and cultures, which was made possible through digital media. Both in the images visible in Figure 5.2 and in the video of indigenous leader Ailton Krenak in Figure 5.3, the student worked with designs available on the Web and, through them, expanded her understanding of the course themes.

Figure 5.2: Introduction of Work at the Pre-publication Stage

Caption: Decolonization of Language Teaching Journal.
Work in progress by student Carolina Dametto: "The Revitalization of Indigenous Language Through the Lens of Decolonization."
With the need to break colonialist thinking, reaffirm ways of coexisting, being, and existing, as well as to value language and culture—since there is no absolute truth—decolonial thought has been exercising the act of resistance to overcome colonialist thought and practices in the country. Decoloniality assumes the role of rescuing the identity of a specific people, as well as re-signifying their culture, way of life, and traditions for the exercise of their identity, which have been suffocated by Eurocentrism."

Source: Screenshot. CGScholar. Available at: https://cgscholar.
com/creator/works/116656/versions/231744?mod=tools&tool-
group=project_toolgroup&tool=status_too.

Figure 5.3: Excerpt from Carolina's Work

Caption: Work in progress by student Carolina on a video featuring indigenous leader and writer Ailton Krenak, who highlights that over 600 native languages of Brazil have already become extinct:

"When discussing the indigenous ethnicity, we can change the concept that the Portuguese language, officially recognized in the country, is far from being the first and only language. To give you an idea, according to IBGE (Brazilian Institute of Geography and Statistics [Portuguese: Instituto Brasileiro de Geografia e Estatística]) statistics, there are over 300 different ethnic groups, and we encounter two linguistic branches, Tupi and Macro-Jê, meaning there are approximately more than 165 indigenous languages spoken in Brazil. It's important to note that many indigenous people are polyglots, as they not only speak their native language but also engage in multilingualism among the tribes. They must speak each other's language to maintain communication, and in addition to multilingualism, they also speak Portuguese. According to the 2010 Census, contrary to what we might think, the most spoken indigenous language is not Guarani or Xavante, but Tikuna, spoken by approximately 33,000 people. In this language, the meaning of a word changes according to the tone of the syllable in which it is spoken."

Source: Screenshot. CGScholar, 2023.

Note: Video shown three years ago on Canal Brasil, program *Cinejornal*, and available with free access on YouTube at https://www.youtube.com/watch?v=Uwk_hgE0vMs.

In other words, to compose her reflection, Carolina worked exclusively within the realm of cyber-social meanings, and within this realm, she was able to utilize alphabetic writing alongside multimodal representations, generating a new academic discourse. In this case, had indigenous cultural elements not been available in digital media, Carolina would not have been able to articulate her reflection in the manner she did, as she did not have direct access to an indigenous culture with the characteristics she presents in her work. Therefore, the course provided the student with the opportunity to appropriate designs dispersed across the Web to compose a more organized reflection on the theme, producing knowledge that can return to the community and be made available on the Web, thus expanding the space for enunciation of the indigenous culture she wishes to defend.

In this way, the cyber-social environment makes it possible to expand the so-called third space,[6] resulting from the contact zones between the world of the colonizer and the worlds of the colonized. In this sense, the digital allows for the rearticulation of silenced identities and histories, challenging and reformulating the narratives imposed by the colonizer. It can thus be a site of resistance where subaltern voices express themselves and create meanings outside colonial paradigms. It is a space of transculturality, where new forms of knowledge and identities can emerge. The "third space" also represents a place of dialogue and synthesis between different cultures and perspectives, promoting a more complex and nuanced understanding of the world. Published on the Web, this knowledge can circulate in a post-national space (Mignolo, 2003) and find new forms of life.

The works carried out in this writing project are interconnected at the intersection of local histories, which students bring, with global networks through the cyber-social learning environment. This, in turn, connects the involved universities and expands them into various communities that use the CGScholar platform to develop knowledge ecologies. We are only beginning to explore the possibilities of an environment with these characteristics for education.

Literacy as a Social Process

I now wish to focus on the idea of literacy as a process. Here, I propose the notion of "process" as a series of temporally distributed actions that involve the interaction of individuals and aim at achieving a particular outcome. From this perspective, becoming literate occurs when a person engages in specific activities

---

[6.] Conforme Bhabha (1998), Mignolo (2003). Ver mais sobre o conceito na Kalantzis et al. (2020, p. 176).

within the literacy process, which requires them to mobilize or develop certain competencies to reach particular goals. Based on this definition, I would like to introduce two additional elements to further this reflection. According to Street (2014, p. 154), "literacy is not limited to the acquisition of content; it is a process. Literacy is acquired in specific contexts and in particular ways, and the learning modalities and students' social relationships with the teacher constitute forms of socialization and acculturation."

In the aforementioned text, the author emphasizes the importance of socialization in the literacy process. For the author, literacy is linked to interaction and the dynamics between individuals and the social groups with which they establish relationships. According to him, without socialization, access to literacy practices is not possible. Researchers Lankshear and Knobel (2011, p. 18) hold a similar view, stating that "to participate effectively and productively in any literacy practice, people need to be socialized into it." Socialization involves assuming a social identity and using language. Since literacy is a social experience, it is ideological, reflecting values, attitudes, and worldviews, as it is embedded within a social context; it is never neutral. Those who advocate for the neutrality of literacy are often seeking to impose their own values as the dominant standard.

Street also points out that there are "forms of acculturation" in students' social relationships with the teacher. This happens because, first and foremost, the teacher is a physical presence. In and of itself, the way a human body is situated in the world expresses meanings. This already places the relationship within the realm of language and values. In a human relationship, people share experiences and worldviews through dialogue, and in this process, acculturation occurs, in the sense that one person may acquire cultural traits from another. The teacher, however, holds a position of authority and serves as a role model for the student, which confers greater responsibility and awareness of how their own values can impact students' learning. This is a natural occurrence and generates differentiation processes that are productive of subjectivity.

Assessment in Literacies: Beyond Testing

Another experimental approach in this study was carried out using the artificial intelligence (AI) system available on the platform utilized. The system recorded the writing data produced by the students throughout the course. Part of these data were processed, structured, and converted into semantically readable visualizations, which were then provided to the students as a source of feedback.

Figure 5.4 shows the recorded data from the writing process in a visualization that allows for the analysis of the different versions of Anna's text, tracking her edits between each version, the comments from her reviewers, and the rubric criteria. Each topic was handled separately through navigation tabs, as seen in Figure 5.4: Diff, Original, Changed, Review 1, Review 2, and Review Criteria.[7] Above these tabs, consolidated information about the writing process from version to version is provided. This figure presents comparison data between version 2 and version 3 of the work, including the project name, the number of reviews received, the percentage of the text edited (in this case, 38.74%), the original length, and the length after edits. Below this information, the text is made available.

The visualization below shows the content available in the Diff tab, which displays the consolidated differences between the text versions. In green, the paragraphs added by the student are highlighted, following the reviewer's recommendations. Two paragraphs down, there are light red dashed highlights representing the sections the author removed during the revision process. Through these visualizations, I followed Anna's thought development during her writing process.

Building on the possibility of recording and documenting the writing produced in the learning environment, I now reflect on the nature of assessment supported by a platform with these characteristics. To do so, I use the test-based assessment paradigm as a basis for comparison. I take this approach because, as a history teacher for over 10 years, I had to prepare high school students for the National High School Exam (ENEM) and college entrance exams. Even today, as a digital literacies instructor, I continue to work with teachers who need to prepare students for these exams and still witness how their logic influences the entire learning process. In addition, it is important to note that public service entrance exams also use tests to evaluate candidates.

According to Horta Neto (2018), in the Brazilian context, cognitive testing was introduced as a systematic method of assessment between the 1960s and 1980s, coinciding with the emergence of large-scale college entrance exams. Starting in 1988, these methods were widely accepted by governments of different ideological orientations as a criterion for evaluating education, becoming an integral part of education systems.

The implementation of external exams in the education system has raised questions and challenges in schools' pedagogical processes. It has been observed that

---

[7.] Estas categorias estão descritas na legenda da Figure 5.4.

Figure 5.4: Consolidated View of the Edits Made in a Work During the Peer Review Process

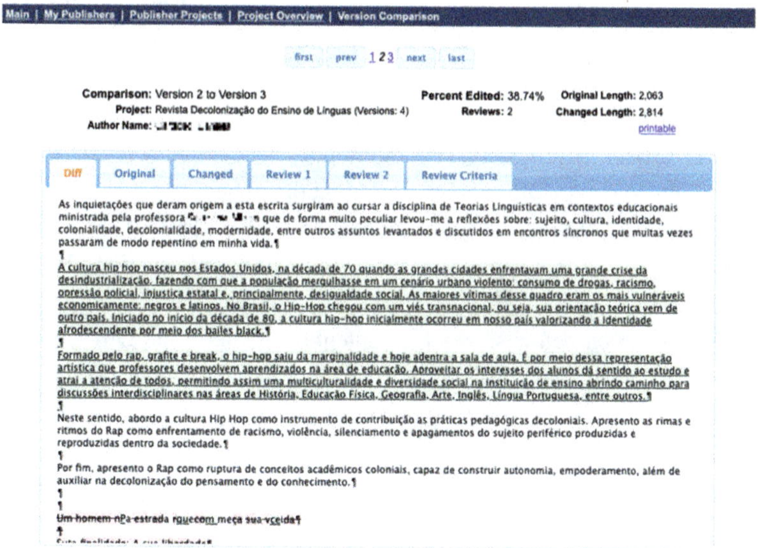

Main | My Publishers | Publisher Projects | Project Overview | Version Comparison

first   prev   1 2 3   next   last

Comparison: Version 2 to Version 3   **Percent Edited:** 38.74%   **Original Length:** 2,063
Project: Revista Decolonização do Ensino de Línguas (Versions: 4)   **Reviews:** 2   **Changed Length:** 2,814
Author Name: ⬛⬛⬛ ⬛⬛⬛   printable

Diff	Original	Changed	Review 1	Review 2	Review Criteria

As inquietações que deram origem a esta escrita surgiram ao cursar a disciplina de Teorias Linguísticas em contextos educacionais ministrada pela professora ⬛ ⬛ ⬛ ⬛ ⬛ que de forma muito peculiar levou-me a reflexões sobre: sujeito, cultura, identidade, colonialidade, decolonialidade, modernidade, entre outros assuntos levantados e discutidos em encontros síncronos que muitas vezes passaram de modo repentino em minha vida.¶
¶
A cultura hip hop nasceu nos Estados Unidos, na década de 70 quando as grandes cidades enfrentavam uma grande crise da desindustrialização, fazendo com que a população mergulhasse em um cenário urbano violento: consumo de drogas, racismo, opressão policial, injustiça estatal e, principalmente, desigualdade social. As maiores vítimas desse quadro eram os mais vulneráveis economicamente: negros e latinos. No Brasil, o Hip-Hop chegou com um viés transnacional, ou seja, sua orientação teórica vem de outro país, iniciado no início da década de 80, a cultura hip-hop inicialmente ocorreu em nosso país valorizando a identidade afrodescendente por meio dos bailes black.¶
¶
Formado pelo rap, grafite e break, o hip-hop saiu da marginalidade e hoje adentra a sala de aula. É por meio dessa representação artística que professores desenvolvem aprendizados na área de educação. Aproveitar os interesses dos alunos dá sentido ao estudo e atrai a atenção de todos, permitindo assim uma multiculturalidade e diversidade social na instituição de ensino abrindo caminho para discussões interdisciplinares nas áreas de História, Educação Física, Geografia, Arte, Inglês, Língua Portuguesa, entre outros.¶
¶
Neste sentido, abordo a cultura Hip Hop como instrumento de contribuição as práticas pedagógicas decoloniais. Apresento as rimas e ritmos do Rap como enfrentamento de racismo, violência, silenciamento e apagamentos do sujeito periférico produzidas e reproduzidas dentro da sociedade.¶
¶
Por fim, apresento o Rap como ruptura de conceitos acadêmicos coloniais, capaz de construir autonomia, empoderamento, além de auxiliar na decolonização do pensamento e do conhecimento.¶
¶
¶
Um homem n̄a estrada roue com meça sua vceida¶
¶
Suta finalidade: A sua liberdade¶

Caption: The image shows the documentation of the writing project. The teacher has access to this type of documentation for each student. The figure displays the set of interactions made during the text creation process throughout the project. At the top of the image, it shows which versions are being compared. Below this section, there is a navigation menu with tabs. The first tab, from left to right, is called "Diff" (Different) and displays the interventions made in the text, line by line. The following tabs allow the user to view each part of the text separately. In "Original," we have the original version submitted by the creator for review; in "Changed," the version modified by the creator based on the feedback received. Next, there is a tab for each review received, showing the scores assigned for each rubric criterion (see Hattie & Timperley, 2007; Smith et al., 2017 for details) by the reviewer, along with comments on each criterion and other comments exchanged between the text creator and the reviewer. Lastly, the "Review Criteria" tab presents all the rubric criteria used in the project, including the name of the criterion, its description, and the corresponding description for each point on the Likert scale.

Source: Screenshot. CGScholar, 2023.

such assessments are limited to quantifying results obtained from cognitive tests, which form the basis of the main instrument for measuring basic education—the Basic Education Development Index (IDEB). This index has become a tool used by the federal government to measure education quality and formulate public policies, an approach that has triggered practices of ranking and competition among schools. Consequently, schools have adopted strategies to improve their scores, such as narrowing the curriculum to topics that will be covered on the exams and pre-selecting students to take the tests. Additionally, schools that do not perform well in these exams may face pressure from families.

Since the administration of Fernando Henrique Cardoso,[8] the Brazilian educational system has been influenced by this logic. In this sense, Horta Neto (2018) emphasizes that financial resources are allocated for the development of tests, to the detriment of other possible educational needs. Educational institutions often find themselves unable to initiate deeper pedagogical debates with their school community. Instead, they are compelled to adapt to external and abstract measures.

The reading and interpretation experience stimulated by the tests assumes an understanding of meanings that have been directly and linearly conceived as something intrinsic to the texts, following the intentions and meanings supposedly assigned by their authors ("what the author intends to say"). Consequently, knowledge and competencies become elements to be demonstrated in assessments as a successfully acquired learning process, "writing correctly" or showing that the "correct" meanings of the texts have been learned by providing the "right answers" in multiple-choice comprehension tests.

When analyzing the logic of tests for assessing reading comprehension, Castell et al. (1986) observe a disconnect between the type of literacy addressed in schools and that which is considered relevant in communities and professional contexts. It is as if the tests become an end in themselves, requiring students to study solely to complete them.

Investigating the possibilities of assessment based on literacy theories, Duboc argues that a new form of evaluation should consider design and include "the subjectivity, interests, intentions, commitments, and purposes of students regarding their learning processes" (2007, p. 110). I believe that we can practice an education with these qualities while simultaneously documenting learning through the multiple records allowed by AI, not only to assign grades and/or scores to students but to reframe their practice. In other words, we can transform what

---

[8.] Em 1998, o Ministério da Educação instituiu o Exame Nacional do Ensino Médio (Enem). [In 1998, the Ministry of Education instituted the National High School Exam (Enem).]

has historically been experienced as burdensome by many students—receiving a grade or score—into something motivating. By doing this, we will initiate a new practice that, in the future, could help to overcome the classification systems underlying tests, engage students in a more constructive logic, and thus expand the opportunities for inclusion that education can provide.

AI-Powered Assessment: A Student-Centered Approach

Recent innovations in the field of educational data mining, focused on data analytics methods that facilitate the documentation and visualization of learning, have been instrumental in diversifying assessment processes. This shift enables a transition from predominantly unidirectional approaches, where teachers assess performance outcomes, to more multidirectional and student-centered methods.[9] Examples include various models of peer assessment, self-assessment, and learning analytics systems, which encourage students to take greater ownership of their learning and assessment processes.[10]

These ideas are not novel in the educational field, as demonstrated by various proposals from authors within the framework of historical-critical pedagogy,[11] who have always advocated for a rebalancing of agency between teachers and students, and for an education aimed at fostering critical thinking and creativity. The central point of the argument is that current technologies provide means to diversify and expand various pedagogical approaches on a larger scale, leveraging the opportunities of the digital environment for reading, writing, and collaboration.

Among the innovations under development, learning analytics technologies stand out. This field is dedicated to measuring, analyzing, and reporting data related to students and their respective educational contexts, with the goal of understanding and improving both learning processes and the environments in which they take place.[12] Such solutions are implemented through AI systems structured based on ontologies—conceptual categories created by humans—which enable the selection of the types of data to be collected, making them semantically readable.[13] The ability to provide detailed insights into students'

[9.] Cope and Kalantzis (2016), Saini (2023).

[10.] de Brún et al. (2022), Martin & Ndoye (2016); Ndoye (2016) *apud* Saini (2023).

[11.] Freire (2005), Saviani (2021), Monte Mor (2019).

[12.] Hernández-de-Menéndez et al. (2022).

[13.] Cope and Kalantzis (2016, 2020).

learning and performance, aiming to enable not only students but also teachers to define more humane and effective educational strategies, has significantly driven research and the practical application of learning analytics in various fields of knowledge (Saini, 2023).

As an example of recent applications of AI in education, Lane et al. (2016) highlight the vast potential of AIED (artificial intelligence in education) in fields such as collaborative, immersive, affective, and exploratory learning. Rosemary Luckin (2018) emphasizes the various ways in which machine learning can complement human learning. A group of 11 experts developed an analytical framework to address the growing integration between AI and human learning (Markauskaite et al., 2022). Meanwhile, the research team led by Cope and Kalantzis, with whom I had the opportunity to collaborate on several projects during my doctoral studies, developed and evaluated AI technologies to support collaborative learning and educational data analytics, providing formative feedback.[14]

However, it is observed that some authors perpetuate the conception of AI as an emulation of the human being (Salas-Pilco et al., 2022). This perspective has led to the resurgence of behaviorist approaches in education, highlighting a symbiosis between the view of AI as a human reflection and behaviorist educational methodologies, as pointed out by Cope and Kalantzis (2023b).

In line with the cyber-social perspective, the originality of the AI system analyzed in this study lies in the fact that it is subordinated to a learning architecture, in relation to which the collected data can enable humans, teachers, and students to construct meaning from them. They are not quantifiers that mark people's subjectivity, turning them into objectives, but rather can represent certain properties of the learning process. This is the subject of the next section.

Integration Between Machine Learning and Semantics

To continue this discussion, I refer the reader to Chapter 1 of this book, where I characterize the foundations of a cyber-social learning environment. I remind you that the term "cyber" here refers to cybernetics as a science of feedback. Thus, one of the proposals presented by the AI-based assessment practice analyzed in this study is to transform each assessment indicator into a source of feedback that the student can activate while learning.

[14.] Cope et al. (2021).

To this end, the learning analytics system used in this study considers as evidence not data representing the student's conformity to an external reference measure, but rather a type of information that reflects what the student has done to learn within the proposed context. The examples I present in this chapter are captured through actions: students wrote texts from different perceptual positions or postures, such as commenting, publishing updates, writing projects, and reviewing; they collaborated with peers; they worked with multimodal writing; they used various digital tools available on the platform; they wrote feedback; they reflected on their own work and documented this reflection in writing. The data reflect these actions and make learning visible (Figure 5.5).

In the dynamic expressed by the figure, what is being assessed was decided by the teacher in dialogue with their students. Each bar in the graph represents an action performed in the learning environment. To the right of the circle, there is a legend containing the names of these actions, or indicators. The categories presented there make the data semantically readable for both students and teachers. This approach represents the paradigm proposed by the Cyber-Social Learning Group for working with AI[15]: It involves combining the computational power of computers with structures of human meaning, or theories. In the case studied, the theory incorporated into the development of the ontologies comes from the pedagogy of multiliteracies. This serves as the epistemic foundation that, in relation to the empirical base represented by the data, results in the possibility of proposing an assessment of multiliteracies for the digital society.

This assessment proposal allows for questioning the test-based model. The essence of the argument is this: if each data point represents an assessment point, why should we limit ourselves to a restricted sampling of tests? Furthermore, the emphasis is that all assessment is formative, distributed in the form of constructive and actionable feedback. This makes summative assessment merely a retrospective window into the learning area that has been traversed and evidenced in the data from formative assessment.[16] In this proposal, AI progressively supports students' learning.

[15.] Group coordinated by professors Bill Cope and Mary Kalantzis, with research related to the development and applications of CGScholar. The reader can learn more at this link: Cyber-Social Learning—New Learning Online: https://newlearningonline.com/e-learning.

[16.] Cope and Kalantzis (2019, p. 538).

Figure 5.5: CGScholar Learning Analytics System: Analytics

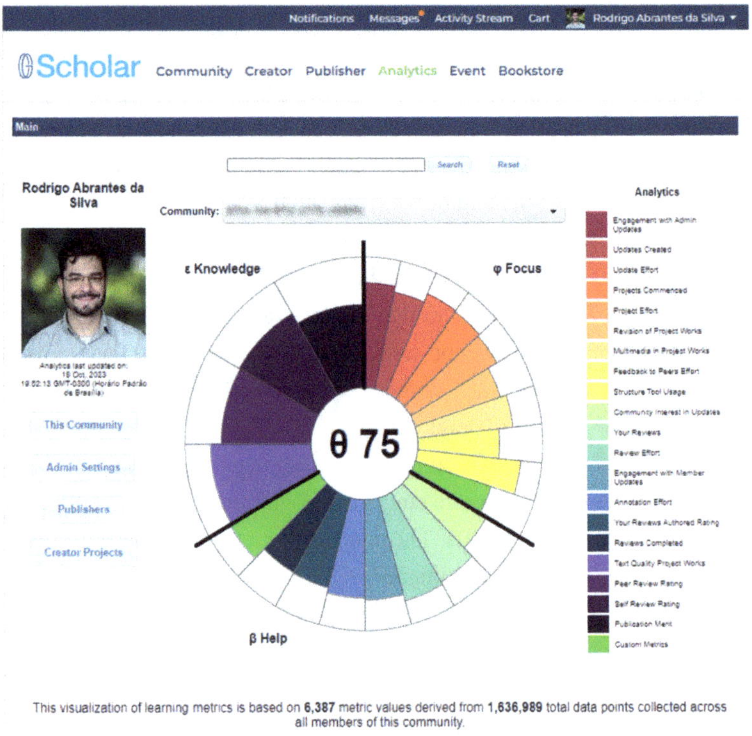

This visualization of learning metrics is based on **6,387** metric values derived from **1,636,989** total data points collected across all members of this community.

Caption: By moving the mouse cursor over each bar represented in the circle, a detailed description of the indicator is displayed, containing the proposed objective, the class average, the student's position, and suggestions on what they can do to improve their performance. Teachers can edit these fields in the tool's settings, as well as add new indicators.

Source: CGScholar. Screenshot, 2023.

Such an approach involves dealing with social complexities that were previously not possible within the communicative structure of the traditional classroom, including peer review and tracking contributions in discussions within virtual learning environments. This materializes the concept of cyber-social learning, which focuses, as we have discussed in this work, on providing recursive feedback, stimulating critical and creative thinking, and developing collaborative intelligence.

Point of Reflection: An Inclusive
Assessment That Values the Student

Traditionally, the reading and interpretation experience underlying tests involves understanding meanings that have been directly and linearly conceived as something intrinsic to the texts, following the intentions and meanings supposedly attributed by their authors ("what the author wants to say"). As a result, knowledge and the acquisition of skills become elements to be demonstrated in assessments as evidence of successful learning, where the "correct" meanings of the texts have been absorbed and are proven through proper writing and "correct answers" in multiple-choice comprehension tests. In current conditions, this traditional conception of literacy presents a rather narrow focus, as it does not stimulate skills that are now considered essential.

The most harmful effect of testing is social segregation (Cope & Kalantzis, 2018), as it divides people by score ranges and conditions access to certain social circles based on these parameters. The points system is internalized by students from the very beginning of their literacy.

Test-based assessment focuses on preparing students for entrance exams, among other selection processes based on tests, thus representing a tautological cycle. The results of these assessments lead teachers to categorize students by levels of achievement at a given moment in the course, in relation to curricular objectives, showing who is ready to advance and who is not, a measure referred to as norm-referenced.[17]

É uma medida que permeia e consolida a desigualdade em todo o sistema educacional. Conforme destacado por Cope and Kalantzis (2018), a base subjacente para a distribuição dos alunos na curva de sino, que é gerada pela avaliação normo-referenciada (ver Figure 5.6), é, de fato, o racismo. Esse sistema naturaliza a hierarquização entre as pessoas, reforçando uma ordem social desigual e injusta, segundo os autores. Adultos, inadvertidamente, corroboram esse comportamento, ao elogiarem efusivamente as crianças que conquistam notas 10. [It is a measure that permeates and consolidates inequality throughout the entire educational system. As highlighted by Cope and Kalantzis (2018), the underlying basis for the distribution of students on the bell curve, which is generated by norm-referenced assessment (see Figure 5.6), is, in fact, racism. According to the authors, this system naturalizes the hierarchy among people,

---

[17.] Kalantzis et al. (2020, p. 375).

Figure 5.6: Norm-Referenced Assessment

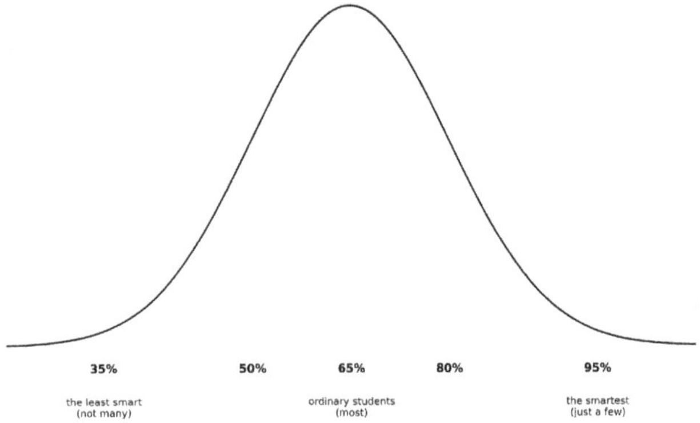

| 35% | 50% | 65% | 80% | 95% |
| the least smart (not many) | | ordinary students (most) | | the smartest (just a few) |

Source: Kalantzis et al., 2020, p. 376.

reinforcing an unequal and unjust social order. Adults inadvertently corroborate this behavior by effusively praising children who achieve perfect 10s.]

In a learning environment governed by the logic of tests, it is forgotten that the human knowledge environment is social. This neglect occurs because tests tend to suppress this dimension by establishing a situation in which learning is defined by individual cognition, anchored in long-term memory. What do we do when we subject students to years of schooling based on this practice, and what kind of individuals do we form?

The so-called 21st-century skills, such as collaboration, creativity, critical analysis, and communication, have been discussed and valued for a long time. However, a crucial challenge arises: how to reconcile the need to develop these essential skills with assessment systems that still rely on cognitive tests, such as entrance exams, the National High School Exam (ENEM), and public competitions? These assessment institutions remain anchored in conventional pedagogical traditions, which, in turn, raises questions about the adequacy of the educational approach in light of contemporary demands.

In this sense, the disparity between the skills promoted by the traditional education system, centered on cognitive assessments, and the skills required by today's world becomes evident. The emphasis on memorization and the reproduction of procedures and content tends to neglect students' ability to work collaboratively, adapt to new situations, and critically analyze information. Thus, a pertinent question arises: what is the role of traditional pedagogy in shaping individuals capable of facing the complex and interconnected challenges that characterize contemporary society?

Although assessment tests continue to be a fundamental tool in educational evaluation, it is essential to consider how they can be adapted or complemented to accommodate the development of essential 21st-century skills. Approaches that stimulate collaboration, creativity, critical analysis, and communication can contribute to a more holistic assessment of students' potential. Furthermore, reflecting on the role of education and assessment in shaping future generations is crucial to ensure that educational systems evolve in accordance with society's changing needs.

Figure 5.7 represents a very common situation in various basic and higher education classrooms: the moment of the test. The judgment. After the grade

Figure 5.7: Traditional Assessment

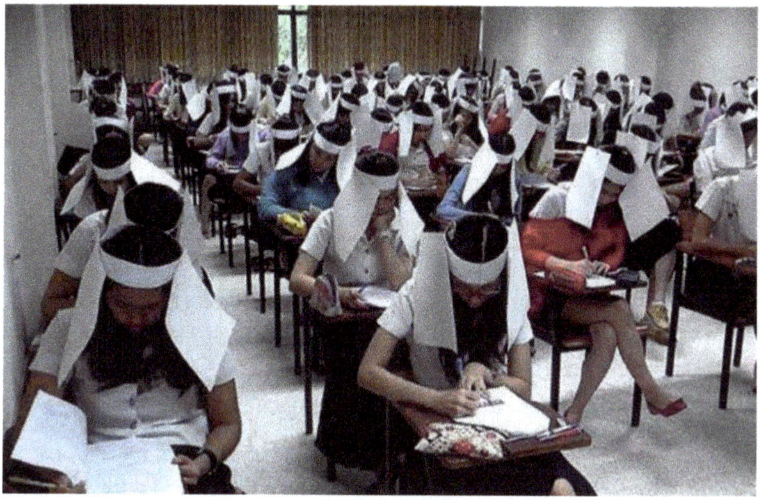

Caption: This image is from an undergraduate course class at a university in Thailand. It was circulated worldwide in 2013. In Brazil, it was published by UOL Educação. Many teachers to whom I show this image react with astonishment. This is because the image starkly reveals, with the "anti-cheating" outfits worn by the students, the inhumane aspect of assessment tests.

Source: UOL Educação. Accessed on August 6, 2023. [Access to the report and image is granted by the UOL Portal.]

is issued, something will happen in the student's subjectivity. I could provide countless similar contemporary images based on my observations in schools. Technological versions of this approach allow for significantly increased control over student behavior, with biometric resources capable of detecting and classifying bodily reactions.

According to Cope and Kalantzis (2019, 2020), taking tests has been a disturbing experience for students, as it prioritizes the interests of educational system managers over the well-being of students. As a teacher, I find this argument very representative of my own experience. I have always observed students feeling stressed before, during, and after a test. Before the test, many feel anxious, trying to guess which content will be covered in the assessment. During the test, many feel that they are not at their best to perform well; some report experiencing a "blank" in their memory and find themselves unable to do something they would have been able to do at another time. After the test, students are anxious to know what grade they received and whether it will be enough to meet the average or other goals. Given that this practice conditions the school calendar, we understand that the school experience becomes reduced to managing situations of stress and competition arising from test orientation.

These issues are of great relevance, but often, in the hectic daily life of basic education teachers with whom I interact, they may seem disconnected due to the overload of tasks, such as creating and grading tests, as well as generating performance reports. In this context, I consider experiences like this a crucial facet of the malaise in education in the 21st century. We often delude ourselves into believing that we can have absolute control and generate an objective measure of learning through practices that, instead of stimulating, often seem to stifle the development of students' potential.

Another problem with tests is that we create the idea that we can assess all students with the same instrument, as if everyone learned in the same way and had the same resources and needs. Again, this idea arises because this evaluative practice is dissociated from the social relationships involved in the production of knowledge. We know that life experience is fundamental to students' learning. Their interactions, social spaces, and access to cultural resources such as museums, libraries, bookstores, theaters, and cinemas form a repertoire of activities that contribute to learning. Thus, the inequality of access to cultural resources is fundamental to understanding the inequality in learning processes. Yet people with different levels of access to cultural resources are assessed by the same instruments and in the same way.

Despite these criticisms, tests continue to be widely used in education, particularly in assessing literacy skills. This is because tests are seen as a cost-effective and efficient way to measure a student's knowledge and understanding of a particular subject. However, it is important to recognize that they create a situation distinct from the learning process itself and do not provide a comprehensive understanding of a student's literacy competencies (Kalantzis et al., 2020).

Tests can also be problematic in situations where students learn a lot but perform poorly on them. In these cases, it is often recommended that students develop strategies, such as learning to manage stress by identifying the sources of their nervousness. However, it is essential to recognize that the problem lies with the instrument itself, not with the student.

Ultimately, it is crucial to acknowledge that tests are a limited tool for assessing literacy skills and should be used alongside other evaluation methods, such as performance-based assessments and portfolios. By adopting a more holistic approach to literacy assessment, educators can gain a broader understanding of their students' competencies and provide the support and resources needed for their success.

The effective development of writing skills is vital for students' growth and academic achievement. Since writing is an essential skill across various disciplines—including sciences, humanities, and technical courses—it is imperative to reevaluate how it has been utilized in these contexts. Cope and Kalantzis (2016) argue that this reevaluation is underway and has led to a shift in the pedagogical focus toward literacy, emphasizing the productive mode of writing over the receptive mode of reading.

In the traditional assessment model, the focus is on testing and classifying students, generating anxiety and stress while reproducing inequalities. In contrast, the cyber-social model I am analyzing emphasizes providing detailed and enriching feedback to students, composed of data that informs their performance. In this process, evidence of learning is represented by data reflecting what each student has accomplished to achieve their goals. All evidence considered for assessment is a moment of the learning experience itself, returning some form of feedback to the student: a comment from a peer, a teacher, or a teaching assistant; a data visualization in Analytics; or an automated machine comment. There are no measures outside the realm of students' experiences, and there is no final exam.

With less emphasis on competition among students and greater appreciation for mastery and student success, this form of assessment can help reduce stress and pave the way for analytical thinking, which is regarded today as the most important skill for the future.

CHAPTER 6

New Learning Journey: Inclusion Through the Cyber-Social Environment

Your theories, large and small, change with new experiences. People who cease to have and look for experiences with new and different people eventually stagnate. Their theories—and, in that sense, too, their brains—stultify, cease to grow in important ways, become less and less "true" in any meaningful and helpful sense. They can become a danger to themselves and others. It is only by seeking new experiences with diverse people that our theories and our worldview can grow toward truth, though we never reach truth as a final destination, but seek it only as a journey.

James Paul Gee (2020)[1]

I bring here a reflection on the possible paths to a new learning journey, based on the theoretical and practical frameworks presented in this book and drawn from my doctoral thesis. We will begin by addressing some structural challenges faced today by educators in their teaching practice and alternatives to overcome them, in a relationship that aligns with the cyber-social paradigm. For this, I take as a backdrop that education is a common good, a public good, as stated in the Brazilian Federal Constitution, Chapter III, Article 205: "Education, a right of all and a duty of the State and the family, shall be promoted and encouraged with the collaboration of society, aiming at the full development of the individual, their preparation for the exercise of citizenship, and their qualification for work."

I emphasize this point because the feasibility of implementing a digital inclusive education within the country's education networks often raises questions due to, among other factors, the lack of material conditions for access to computers and the internet. Instead of approaching this discussion from a statistical perspective,

---

[1] James Paul Gee (2020).

I propose to do so based on the autoethnographic account of Prof. Dr. Ana Paula Guimarães,[2] who teaches English in a municipal elementary school in Cidade Tiradentes, a neighborhood in the East Zone of São Paulo.

The teacher's account presents students who have only one cell phone for studying, which must be shared among parents and siblings, and/or rely on the neighbor's internet signal to access the platform used by the school. Furthermore, she highlights that owning computers and cell phones is not a priority for these students:

> *A large part of this student body is primarily motivated to attend school by the meals they receive, which are increasingly scarce and meager, served twice per period. Many children arrive at school dirty, disheveled, and wearing flip-flops that are missing pieces or pieced together from different old flip-flops. (Guimarães, 2023, p. 87)*

As if this material deprivation were not adverse enough, it is further exacerbated, according to the teacher, by underlying values in the school curriculum that lead children to internalize notions that position them as inferior members of society. Such notions operate through mechanisms of hierarchization. An illustrative example highlighted by the author is the derogatory perception of the group regarding their linguistic abilities, exemplified by the phrase, which students internalize very early on: "You can't even speak Portuguese; how are you going to speak English?"

This expression not only underestimates the linguistic competence of children in their mother tongue but also establishes a hierarchy between Portuguese and English, implicitly valuing mastery of the latter more highly. Such an attitude reflects a devaluation of the local language and culture, while also perpetuating barriers to the academic and personal development of these students, who are already situated in a context of socioeconomic disadvantage. Today, the English language is a mandatory component of the Brazilian Basic Education curriculum, and its teaching is related to the cultures of the countries that have built the current regime of globalization. In this sense, it is essential to question what the conditions are for these students' participation in this system.

With globalization came various trends that were supposedly intended to benefit everyone: the knowledge and creativity economy, a culture of participation, flexible

---

[2] Her doctoral thesis: "Emergency Pedagogy with the Challenges and Charms of English Language Education in a Public School in Cidade Tiradentes: Multiliteracies, New Literacies, Translanguaging, and Ubiquity". I followed this professor's journey during the years in which I conducted my doctoral study.

work, creative leisure, the dismantling of vertical and authoritarian hierarchies, freedom of movement, and so on. However, these trends emerged alongside a set of prerequisites that individuals must possess in order to effectively participate in this society: fluency in English, digital literacy, teamwork skills, resilience, flexibility, and so forth. These prerequisites are unevenly distributed because they depend on what Bourdieu termed cultural capital.[3]

The professor's account highlights the experience of a population deprived of the cultural capital required by the market to develop the necessary prerequisites satisfactorily. Although the educator-author aims to work with an education that encourages her students to cultivate the dispositions needed to integrate economically and socially into globalization, her efforts are inhibited at various levels of the education system, which, for her, are designed "not to work."

She cites as an example a project for distributing tablets to students, which were provided with various resources blocked and under a loan agreement that discouraged families from supporting their children's use of them.

In these conditions, the prerequisites of globalization create a significant gap between those who can access different cultural capitals and the resources associated with them and those who are left only with the expectation of having them. "Not everyone participates in the benefits of globalization," the professor emphasizes in the report.

In light of this scenario, I propose an education that promotes equity, "gradually reducing the extensive gap that exists between students in public schools and those in elite schools in São Paulo," as global connectivity and interdependence are

---

[3.] The French sociologist Pierre Bourdieu developed the concept of "cultural capital" in his book *Distinction: A Social Critique of the Judgment of Taste* (Jorge Zahar Editora, 2011). The work presents part of his theory on forms of power in society, particularly concerning the reproduction of social inequalities. For him, cultural capital can be understood in three main forms:

- Incorporated—This form of cultural capital refers to skills, knowledge, ways of speaking, and thinking that a person acquires throughout life, especially during childhood. It becomes part of the individual and cannot be separated from them. For example, linguistic competencies, artistic or literary knowledge, and even the ability to appreciate certain cultural forms. The incorporation of this capital primarily occurs within the family unit and its surroundings, through the cultural habits that the child observes and experiences.
- Objectified—This encompasses physical cultural goods, such as books, works of art, and musical instruments. Although these objects can be transmitted or sold, the ability to appreciate or use them appropriately (the incorporated cultural capital) is not transferable through the physical object. Therefore, simply having books and/or computers is not enough; it is necessary to have access to ways of interacting with these goods so that the values associated with them can be grasped.
- Institutionalized—This refers to the formal recognition of an individual's education and qualifications. A clear example is an academic diploma. It confers a certain status and cultural legitimacy to a person's knowledge and skills, often influencing their career opportunities and social standing.

Bourdieu argues that cultural capital, in its various forms, plays a crucial role in maintaining structures of power and privilege in society. He highlights how upper-class families transmit cultural capital to their children, giving them a significant advantage in terms of access to educational and professional opportunities.

now inevitable, facts of history (Rizvi, 2022, p. 2017). Therefore, it is essential to seek answers on how globalization can be redefined and what the terms of our global interdependence should be, both in the material structure of power and in the way we interpret and represent the world. A new rationality is needed, emphasizing ethical perspectives in national pursuits and global exchanges.

Indeed, adapting to the discourses and materialities of digital technologies represents a prerequisite for participation in various cyber-social spheres—a vital skill for human regulation in different environments. Thus, I consider it essential to reflect on how this adaptation impacts human agency in light of technological development and, more importantly, to have a strategy that reinserts the human element into technology, indicating other possible pathways for these relationships.

Reviewing Our Practical Possibilities

This book presents the cyber-social paradigm as a possible pathway for exercising an education that responds to the dynamics of digital society, from a human perspective rather than a machine-based one. I would like to remind you that the term "cyber" does not refer (only) to technology, but to a recursive relationship, a permanent feedback loop between humans and machines, facilitated by feedback mechanisms. The term "social," in turn, situates the human within this relationship. This perspective allows us to view the digital as a field of social practice and connect it to sociocultural theories of literacy. It also enables us to engage with the repertoire of words and human meanings as inseparable components of digital media, which are always activated and fed back into the system—something I consider important as an educator.

In other words, if we create space for the diversity of human experiences in the digital environment, education will have the potential to expand learning in many ways. This could occur, for example, through collaborative forms of learning and knowledge production in formal education settings, as well as through more horizontal relationships between teachers and students. Transforming the conventional classroom discourse will be a challenge, but it could offer a fascinating opportunity for the reinvention of education.

It is essential to recognize that conventional classroom discourse was effective when learning resources were scarce and depended solely on the spoken word and instructions from the teacher. However, in digital society, we have knowledge, resources, and intelligence distributed across the Web, and each of us can also use our infrastructure to create and distribute resources. Once distributed and readily accessible, a vast stockpile of time and resources can be liberated, rendering the traditional learning model tied to a teacher's speech obsolete.

Similarly, it is evident that there is no need for all students to follow the same schedule, such as classes with fixed time, space, and pace. Engagement with the course does not depend on the simultaneous presence of everyone at the same time and place. The flexibility offered by the Web allows students to learn anytime and anywhere. True engagement is fostered by elements such as continuous feedback, the provision of personalized approaches, collaboration, and recognition of students' efforts by the teacher.

When I began my doctoral study, I questioned whether it would be feasible to create a modern platform based on pedagogical principles without having to adapt or rely on market systems. The CGScholar platform, as it functions so far, shows that it is indeed possible. Its development involved decades of intense and focused work by Professors Dr. Bill Cope and Dr. Mary Kalantzis. As they themselves acknowledge, over the years, many software developers have joined this project after working in a technology corporation, from which they left motivated by the desire to make a difference. These projects require funding from agencies focused on promoting diverse and inclusive education. Therefore, with this book, I aim to signal pathways that can make this new learning journey possible.

Many professions have changed significantly in recent years due to technology, but in education, practices remain the same. This is due to the way it is conceptualized. The hegemonic approaches surrounding technology integration projects in education reproduce the numerous problems already pointed out in Chapter 1 and the opening of this chapter. As Knobel and Kalman (2016) observe, these are projects that focus on acquiring hardware and software and teaching teachers to use them through courses, usually of short duration and taught by outside experts.

As my study indicates, these projects operate with approaches to digital literacy in the singular, that is, an autonomous system of literacy that simply reproduces conventional pedagogical models, without the possibility of effectively emancipating the condition inherited by the students.

On the other hand, the very logic of the binary notation system to represent meanings through computers has opened the field for the emergence of new perspectives on literacy. This event affects various societies with different orientations in time and space. Through it, the records generated throughout history have been digitized, expanding the possibilities for their appropriation and use. Furthermore, the expansion of possibilities to represent and record human meanings has generated real-time recordings of various aspects of life. The repositories and recordings of meanings in digital environments become repertoires for design work (i.e., for constructing meanings). In this dimension, numerous literacy practices, in the plural, are developed.

Thus, I see two layers enhancing multiliteracies:

(1) The first layer consists of the moment of digitization, when new records are created. Programmers operate in this layer. There is no entry or participation in this layer without handling machine languages. This layer is open to study. Linguists can access it by engaging with programmers and studying how they have appropriated and applied languages. There is a vast open research field that can help to understand various aspects of doing on the Web from this perspective.

(2) The second layer consists of the use that people make of graphical interfaces. It is on this layer that studies of digital literacy focus. These literacies occur at the confluence of a technical dimension and a new *ethos* (Lankshear & Knobel, 2011). The technical dimension refers both to the fact that they are based on a technical infrastructure and to the skills needed to navigate the Web through software.

It happens that, to be proficient in this environment and to participate effectively in it, one must be in a position to appropriate literacy and use it contextually. This is what occurred when the Cyber-Social Learning Group appropriated the generative AI of ChatGPT and adapted it to make its pedagogical use relevant in the context of the LDL program courses (see Chapter 3), while also providing an experience through which students could interpret this phenomenon.

Throughout this book, we explore how a cyber-social learning environment allows us to shape new ways of teaching and learning, highlighting the importance of a humanized and ethical approach to technology integration.

The chapters presented offer a reflection on the need for new pedagogical practices capable of transitioning from a conventional model to a more collaborative, interactive, and dynamic learning environment. The incorporation of digital platforms, such as CGScholar, exemplifies how technology can be used to promote a more inclusive and participatory education, where students are not mere receivers of information but co-creators of knowledge.

The cyber-social concept applied to education invites us to see the classroom as an interconnected system, where each element—whether human or technological—plays a crucial role in the learning process. This understanding allows us to recognize the importance of relationships and interactions within this ecosystem, fostering a culture of collaboration driven by continuous feedback and adaptation, essential elements for developing an effective learning environment. The interaction between teachers and students, mediated by digital technologies, redefines the educational dynamic, promoting a more personalized learning experience.

BIBLIOGRAPHY

Abrantes da Silva, R., & Mizan, S. (2022). Decolonial practices on the educational platform CGScholar: Subjectification, ecology of knowledges, and the design of rhizomatic multimodal texts. *Ubiquitous Learning: An International Journal, 15*(2), 19–35. https://doi.org/10.18848/1835-9795/CGP/v15i02/19-35; https://cgscholar. com/bookstore/works/decolonial-practices-on-the-educational-platform-cgscholar

Adams, T. E., Jones, S. H., & Ellis, C. (2015). *Autoetnography: Understanding qualitative research*. Oxford University Press.

André, M. (2013). *Etnografia da prática escolar* [Ethnography of school practice]. Papirus Editora.

Barton, D., & Lee, C. (2015). *Linguagem online: Textos e práticas digitais* [Online language: Texts and digital practices]. Parábola Editorial.

Berners-Lee, T., Hendler, J., & Lassila, O. (2001). The semantic web: A new form of web content that is meaningful to computers will unleash a revolution of new possibilities. *Scientific American, 284*(5), 29–37. https://www-sop.inria. fr/acacia/cours/essi2006/Scientific%20American_%20Feature%20Article_%20 The%20Semantic%20Web_%20May%202001.pdf

Bhabha, H. K. (1998). *O local da cultura* [The place of culture]. Editora UFMG.

Bourdieu, Pierre (2011). *Distinction: A social critique of the judgment of taste*. Jorge Zahar Editora.

Carson, J. G., & Nelson, G. L. (1996). Chinese students' perceptions of ESL peer response group interaction. *Journal of Second Language Writing, 5*(1), 1–19.

Castell, S., Luke, B., & MacLemann, D. (1986). On defining literacy. In S. Castell, B. Luke, & K. Egan (Eds.), *Literacy, society, and schooling: A reader* (pp. 3–14). Cambridge University Press.

Cesarino, L. (2022). *O mundo do avesso: Verdade e política na era digital* [The upside down world: Truth and politics in the digital age]. Ubu.

Chapelle, F. H. (2014). The history and practice of peer review. *Groundwater*, *52*(1), 1. https://doi.org/10.1111/gwat.12139

Cope, B., & Kalantzis, M. (Eds.). (2000). *Multiliteracies: Literacy learning and the design of social futures* (pp. 9–37). Routledge.

Cope, W. W., & Kalantzis, M. (2009). Signs of epistemic disruption: Transformations in the knowledge system of the academic journal. *First Monday, 14*(4), 13–61. https://doi.org/10.5210/fm.v14i4.2309

Cope, B., & Kalantzis, M. (2013). Towards a new learning: The scholar social knowledge workspace, in theory and practice. *E-Learning and Digital Media*, *10*(4), 332–356. https://doi.org/10.2304/elea.2013.10.4.332

Cope, B., & Kalantzis, M. (2016). Big data comes to school: Implications for learning, assessment, and research. *AERA Open*, *2*(2), 1–19. April. https://doi.org/10.1177/2332858416641907

Cope, B., & Kalantzis, M. (2017). Conceptualizing e-learning. In B. Cope & M. Kalantzis (Eds.), *e-Learning ecologies: Principles for new learning and assessment* (pp. 1–45). Routledge.

Cope, B., & Kalantzis, M. (2018). *New learning: Transformative designs for pedagogy and assessment*. Retrieved from https://newlearningonline.com/new-learning

Cope, B., & Kalantzis, M. (2019). Education 2.0: Artificial intelligence and the end of the test. *Beijing International Review of Education*, *1*(2–3), 528–543. http://doi.org/10.1163/25902539-00102009

Cope, B., & Kalantzis, M. (2020a). Futures for research in education. *Educational Philosophy and Theory*, *54*(11), 1732–1739. https://doi.org/10.1080/00131857.2020.1824781

Cope, B., & Kalantzis, M. (2020b). *Making sense: Reference, agency, and structure in a grammar of multimodal meaning*. Cambridge University Press.

Cope, B., & Kalantzis, M. (2023a). On cyber-social meaning: The clause, revised. *The International Journal of Communication and Linguistic Studies*, *21*(2), 1–18. http://doi.org/10.18848/2327-7882/CGP/v21i02/1-18

Cope, B., & Kalantzis, M. (2023b). On cyber-social learning: A critique of artificial intelligence in education. In T. Kourkoulou, A. Tzirides, B. Cope, & M. Kalantzis (Eds.), *Trust and inclusion in AI-mediated education: Where human learning meets learning machines* (pp. 3–34). Springer.

Cope, B., Kalantzis, M. (2023c). Creating a different kind of learning management system: The CGScholar experiment. In M. Montebello (Ed.), *Promoting next-generation learning environments through CGScholar* (pp. 1–18). IGI Global. https://doi.org/10.4018/978-1-6684-5124-3.ch001

Cope, B., Kalantzis, M., & Lankshear, C. (2005). A contemporary project: An interview. *E-Learning and Digital Media, 2*(2), 192–207. https://doi.org/10.2304/elea.2005.2.2.6

Cope, B., Kalantzis, M., & Magee, L. (2011). *Towards a semantic web: Connecting knowledge in academic research.* Chandon Publishing.

Cope, B., Kalantzis, M., & Searsmith, D. (2021). Artificial intelligence for education: Knowledge and its assessment in AI-enabled learning ecologies. *Educational Philosophy and Theory, 53*(12), 1229–1245. https://doi.org/10.1080/00131857.2020.1728732

Cope, B., & Kalantzis, M. (2023). Towards education justice. Em G. C. Zapata, M. Kalantzis, & B. Cope, *Multiliteracies in international educational contexts* (1st ed., pp. 1–33). Routledge. https://doi.org/10.4324/9781003349662-1

da Silva, R. A. (2020). *A plataforma Scholar e o Projeto Piloto USP-UIUC (Universidade de São Paulo e Universidade de Illinois em Urbana-Champaign): Inovações em formação de professores)* [The scholar platform and the USP-UIUC pilot project (University of São Paulo and University of Illinois at Urbana-Champaign): Innovations in teacher training]. *Papéis, 24*(47), 27–50. https://doi.org/10.35699/9784

De Brún, A., Rogers, L., Drury, A., & Gilmore, B. (2022). Evaluation of a formative peer assessment in research methods teaching using an online platform: A mixed methods pre-post study. *Nurse Education Today, 108*, 105166. https://doi.org/10.1016/j.nedt.2021.105166

Duboc, A. P. M. (2007). *A questão da avaliação da aprendizagem de língua inglesa segundo as teorias de letramentos* [The issue of assessing English language learning according to literacy theories]. (Dissertação de Mestrado).

Faculdade de Filosofia, Letras e Ciências Humanas, Universidade de São Paulo, São Paulo, Brasil.

Francis, K. L. (2021). *Sense of community and peer review: A case study of a doctoral dissertation experience* (Doctoral dissertation, University of Illinois Urbana-Champaign). University of Illinois Urbana-Champaign.

Freire, P. (2005). Pedagogy of the oppressed. Continuum. (Original work published 1970)

Gee, J. P. (2013). *The anti-education era: Creating smarter students through digital learning*. St. Martin's Press.

Gee, J. P. (2015). *Literacy and education*. Routledge.

Gee, J. P. (2020). *What is a human? Language, mind, and culture*. Palgrave Macmillan.

Gee, J. P., & Hayes, E. R. (2011). *Language and learning in the digital age*. Routledge.

Gibson, J. J. (1979/2015). *The ecological approach to visual perception: Classic edition*. Psychology Press.

Goldin, I., Narciss, S., Foltz, P., & Bauer, M. (2017). New directions in formative feedback in interactive learning environments. *International Journal of Artificial Intelligence in Education, 27*(3), 385–392. https://doi.org/10.1007/s40593-016-0135-7

Guimarães, A. P. (2023). *Pedagogia emergencial com os desafios e encantos da educação em Língua Inglesa em uma escola pública da Cidade Tiradentes: Multiletramentos, novos letramentos, translanguaging e ubiquidade* [Emergency pedagogy with the challenges and charms of English language education in a public school in Cidade Tiradentes: Multiliteracies, new literacies, translanguaging, and ubiquity]. (Tese de Doutorado). Faculdade de Filosofia, Letras e Ciências Humanas, Universidade de São Paulo, São Paulo, Brasil. https://doi.org/10.11606/T.8.2023.tde-01082023-185942

Gusso, H. L., Archer, A. B., Luiz, F. B., Sahão, F. T., Luca, G. G. de ., Henklain, M. H. O., Panosso, M. G., Kienen, N., Beltramello, O., & Gonçalves, V. M.. (2020). ENSINO SUPERIOR EM TEMPOS DE PANDEMIA: DIRETRIZES À GESTÃO UNIVERSITÁRIA. *Educação & Sociedade, 41*, e238957. https://doi.org/10.1590/ES.238957

Gutiérrez, K. D., & Jurow, A. S. (2016). Social design experiments: Toward equity by design. *Journal of the Learning Sciences, 25*(4), 565–598. https://doi.org/10.1080/10508406.2016.1204548

Haren, R., Harroun, J., & Moraine, K. (2019). CGScholar's analytics: Progress to mastery learning. *Ubiquitous Learning: An International Journal, 12*(2), 1–24.

Hattie, J., & Timperley, H. (2007). The power of feedback. *Review of Educational Research, 77*(1), 81–112. https://doi.org/10.3102/003465430298487

Hernández-de-Menéndez, M., Morales-Menéndez, R., Escobar, C. A., & Ramírez Mendoza, R. A. (2022). Learning analytics: State of the art. *International Journal on Interactive Design and Manufacturing, 16*, 1209–1230. https://doi.org/10.1007/s12008-022-00930-0

Hogan, A. (2020). The semantic web: Two decades on. *Semantic Web, 11*(1), 169–185. https://doi.org/10.3233/SW-190387

Hooper, M. (2019). Scholarly review, old and new. *Journal of Scholarly Publishing, 51*(1), 63–75. https://doi.org/10.3138/jsp.51.1.04

Horta Neto, J. L. (2018). Avaliação educacional no Brasil para além dos testes cognitivos [Educational assessment in Brazil beyond cognitive tests]. *Revista de Educação PUC-Campinas, 23*(1), 37–53. https://doi.org/10.24220/2318-0870v23n1a3990

Hui, Y. (Ed.). (2024). Why cybernetics now? In *Cybernetics for the 21st century* (Vol. 1: Epistemological reconstruction). Hanart Press.

Jones, R. D., & Hafner, C. A. (2021). *Understanding digital literacies: A practical introduction.* Routledge.

Joordens, S., Paré, D., Walker, R., Hewitt, J., & Brett, C. (2019). *Scaling the development and measurement of transferable skills: Assessing the potential of rubric scoring in the context of peer assessment.* Higher Education Quality Council of Ontario.

Kalantzis, M. (2023). Multiliteracies: Life of an idea. *International Journal of Literacies, 30*(2), 17–89. https://cgscholar.com/bookstore/works/multiliteracies-life-of-an-idea?category_id=cgrn&path=cgrn%2F242%2F250

Kalantzis, M., & Cope, B. (2009). *Ubiquitous Learning.* University of Illinois Press. https://www.press.uillinois.edu/books/?id=p076800

Kalantzis, M., & Cope, B. (2016). Learner differences in theory and practice[†]. *Open Review of Educational Research, 3*(1), 85–132. https://doi.org/10.1080/23265507.2016.1164616

Kalantzis, M., & Cope, B. (2020). *Adding sense: Context and interest in a grammar of multimodal meaning* (1st ed.). Cambridge University Press. https://doi.org/10.1017/9781108862059

Kalantzis, M., Cope, B., & Pinheiro, P. (2020). *Letramentos* [Literacies]. Editora da Unicamp.

Kaufaman, J. H., & Schunn, C. D. (2011). Students' perceptions about peer assessment for writing: Their origin and impact on revision work. *Instructional Science, 39*(3), 387–406. https://doi.org/10.1007/s11251-010-9133-6

Keh, C. L. (1990). Feedback in the writing process: A model and methods for implementation. *ELT Journal, 44*(4), 294–304. https://doi.org/10.1093/elt/44.4.294

Knobel, M., & Kalman, J. (2016). Teacher learning, digital technologies and new literacies. In M. Knobel & J. Kalman (Eds.), *New literacies and teacher learning: Professional development and the digital turn* (pp. 1–20). Peter Lang.

Knobel, M., & Lankshear, C. (2015). Digital literacy and digital literacies. *Nordic Journal of Digital Literacy, 2015*, 8–20. https://core.ac.uk/download/pdf/303779381.pdf

Kurzweil, R. (2000). *The age of spiritual machines: When computers exceed human intelligence.* Penguin Books.

Kurzweil, R. (2015). *Como criar uma mente: Os segredos do pensamento humano* [How to create a mind: The secrets of human thought]. Editora Aleph.

Lacan, J. (2007). *D'un discours qui ne serait pas du semblante* [From a discourse that would not be a semblance]. Éditions du Seuil.

Lane, H. C., McCalla, G., Loo, C.-K., & Bull, S. (2016). Preface to the IJAIED 25th anniversary issue, part 2: The next 25 years: How advanced interactive learning technologies will change the world. *International Journal of Artificial Intelligence in Education, 26*(2), 539–543.

Lankshear, C., & Knobel, M. (2011). *New literacies: Everyday practices and social learning.* Open University Press.

Luckin, R. (2018). *Machine learning and human intelligence: The future of education in the 21st century.* UCL Institute of Education Press.

Malerba, J. (2020). *Brasil Em Projetos: História Dos Sucessos Políticos E Planos De Melhoramento Do Reino: Da Ilustração portuguesa à Independência do Brasil.* [History of the kingdom's political successes and improvement plans: From the Portuguese enlightenment to the independence of Brazil]. FGV.

Markauskaite, L., Marrone, R., Poquet, O., Knight, S., Martinez-Maldonado, R., Howard, S., Tondeur, J., De Laat, M., Buckingham Shum, S., Gašević, D., & Siemens, G. (2022). Rethinking the entwinement between artificial intelligence and human learning: What capabilities do learners need for a world with AI? *Computers and Education: Artificial Intelligence, 3*, 100056. https://doi.org/10.1016/j.caeai.2022.100056

Martin, F., & Ndoye, A. (2016). Using learning analytics to assess student learning in online courses. *Journal of University Teaching & Learning Practice, 13*(3), 110–130. https://doi.org/10.53761/1.13.3.7

Mayrink, M. F., Albuquerque-Costa, H., & Ferraz, D. (2021). Remote language teaching in the pandemic context at the University of São Paulo, Brazil. In N. Radić, A. Atabekova, M. Freddi, & J. Schmeid (Eds.), *The world universities' response to COVID-19: Remote online language teaching* (pp. 125–137). [S. l.]: Research-publishing.net. https://doi.org/10.14705/rpnet.2021.52.1268

McGroarty, M. E., & Zhu, W. (1997). Triangulation in classroom research: A study of peer revision. *Language Learning, 47*(1), 1–43. https://doi.org/10.1111/0023-8333.11997001

Meinck, S., Fraillon, J., & Strietholt, R. (2022). *The impact of the COVID-19 pandemic on education: International evidence from the responses to educational disruption survey (REDS).* [S. l.]: UNESCO. https://unesdoc.unesco.org/ark:/48223/pf0000380398

Meyers, B. (2004). *Peer review software: Has it made a mark on the world of scholarly journals?* Aries Systems.

Mignolo, W. D. (2003). *Histórias locais—projetos globais: Colonialidade, saberes subalternos e pensamento liminar* [Local histories—Global projects: Coloniality, subaltern knowledges, and liminal thinking]. Editora UFMG.

Monte Mor, W. (2015). Learning by design: Reconstructing knowledge processes in teaching and learning practices. In B. Cope & M. Kalantzis (Eds.), *A pedagogy of multiliteracies* (pp. 186–209). Palgrave Macmillan. https://doi.org/10.1057/9781137539724_11

Monte Mor, W. (2017). Sociedade da escrita e sociedade digital: Línguas e linguagens em revisão [Writing society and digital society: Languages and linguistics under review]. In W. Monte Mor & N. H. Takari (Eds.), *Construções de sentido e letramento digital crítico na área de línguas/linguagens* [Meaning-making and critical digital literacy in the field of languages]. (pp. 267–286). Pontes Editores.

Monte Mor, W. (2019). Formação Docente e Educação Linguística: uma perspectiva linguístico-cultural-educacional. In W. M. da Silve, W. R. da Silve, & D. M. Campos (Orgs.), Desafios da Formação de Professores na Linguística Aplicada (pp. 187–206). Ed. Pontes.

Monte Mor, W., Duboc, A. P., & Ferraz, D. (2021). Critical literacies made in Brazil. In J. Z. Pandya, R. A. Mora, J. H. Alford, N. A. Golden, & R. S. de Roock (Eds.), *The critical literacies handbook* (pp. 133–142). Routledge.

Monte Mor, W., Gattolin, S. R. B., & Colombo Gomes, G. (2024). Uma conversa sobre formação docente na área de línguas [A conversation about teacher training in the field of languages]. *Entrevista Revista E-SCRITA*, *15*, 222–230.

Montebello, M., Pinheiro, P., Cope, B., & Kalantzis, M. (2018). The impact of the peer review process evolution on learner performance in e-learning environments. In *L@S'18: Proceedings of the fifth annual ACM conference on learning at scale*, *35*, 1–3. https://doi.org/10.1145/3231644.3231693

Moxham, N., & Fyfe, A. (2018). The Royal Society and the prehistory of peer review, 1665–1965. *The Historical Journal*, *61*(4), 863–889. https://doi.org/10.1017/S0018246X17000334

New London Group. (2000). A pedagogy of multiliteracies: Designing social futures. In B. Cope & M. Kalantzis (Eds.), *Multiliteracies: Literacy learning and the design of social futures* (pp. 9–37). Routledge.

Nicolelis, Miguel. 2020. *The True Creator of Everything: How the Human Brain Shaped the Universe as We Know It*. New Haven: Yale University Press.

Oldenburg, R. (1989). *The great good place: Cafes, coffee shops, bookstores, bars, hair salons, and other hangouts at the heart of a community* (2nd ed.). Marlowe.

Oliveira, F. I. [UNESP], & Rodrigues, S. T. [UNESP]. (2014). *Affordances: A relação entre agente e ambiente* [Affordances: The relationship between agent and environment]. Editora Unesp. http://hdl.handle.net/11449/113731

OpenAI. (2022). *Introducing ChatGPT.* https://openai.com/blog/chatgpt

Peters, M. A., Rizvi, F., McCulloch, G., Gibbs, P., Gorur, R., Hong, M., & Hwang, Y. et al. (2020). Reimagining the new pedagogical possibilities for universities post-COVID-19: An EPAT collective project. *Educational Philosophy and Theory, 54*(6), 717–760. https://doi.org/10.1080/00131857.2020.1777655

Pinheiro, P. (2020). Text revision practices in an e-learning environment: Fostering the learning by design perspective. *Innovation in Language Learning and Teaching, 14*(1), 37–50. https://doi.org/10.1080/17501229.2018.1482902

Prensky, M. R. (2010). *Teaching digital natives: Partnering for real learning.* Corwin.

Prensky, M. (2012). *Brain gain by Prensky, Marc.* Palgrave Macmillan.

Rizvi, F. (2022). Education and the politics of anti-globalization. In F. Rizvi, B. Lingard, & R. Rinne (Eds.), *Reimagining globalization and education* (1st ed.). Routledge.

Robinson, K. (2021). *O elemento* [The element]. Lua de Papel.

Saini, A. K. (2023). Assessment and analysis through CGScholar. In M. Montebello (Eds.), *Advances in educational technologies and instructional design* (pp. 38–60). IGI Global. https://doi.org/10.4018/978-1-6684-5124-3.ch003

Smith, D. A. (2017). Collaborative peer feedback. In *International association for development of the information society.* International Association for the Development of the Information Society. https://eric.ed.gov/?id=ED579292

Salas-Pilco, S., Xiao, K., & Hu, X. (2022). Artificial intelligence and learning analytics in teacher education: A systematic review. *Education Sciences, 12*(8), 569. https://doi.org/10.3390/educsci12080569

Saviani, D. (2007). História das ideias pedagógicas no Brasil [History of pedagogical ideas in Brazil] (7ª ed.). Autores Associados.

Shirky, C. (2014). *Here comes everybody: The power of organizing without organizations.* Penguin Books.

Stanley, J. (1992). Coaching student writers to be effective peer evaluators. *Journal of Second Language Writing, 1*(3), 217–233. https://doi.org/10.1016/1060-3743(92)90004-9

Street, B. (1984/2014). *Letramentos sociais* [Social literacies]. Parábola Editorial.

Tennant, J. P., Dugan, J. M., Graziotin, D., Jacques, D. C., Waldner, F., Mietchen, D., Elkhatib, Y., Collister, L. B., Pikas, C. K., Crick, T., Masuzzo, P., Caravaggi, A., Berg, D. R., Niemeyer, K. E., Ross-Hellauer, T., Mannheimer, S., Rigling, L., Katz, D. S., Greshake Tzovaras, B., ... Colomb, J. (2017). A multi-disciplinary perspective on emergent and future innovations in peer review. *F1000Research*, *6*(1151) 1–66. https://doi.org/10.12688/f1000research.12037.3

Tokarczuk, O. (2023). *Escrever é muito perigoso* [Writing is very dangerous]. Todavia.

Turing, A. M. (1950). Computing machinery and intelligence. *Mind*, *59*, 433–460.

Twidale, M., & Hansen, P. (2019). Agile research. *First Monday*, *23*(7). https://doi.org/10.5210/fm.v24i7.9424

Tzirides, A. O., Saini, A. K., Cope, B., Kalantzis, M., & Searsmith, D. (2023). Cyber-social research: Emerging paradigms for interventionist education research in the postdigital era. In P. Jandrić, A. Mackenzie, & J. Knox (Eds.), *Constructing postdigital research* (pp. 85–102). Springer. https://doi.org/10.1007/978-3-031-35411-3_5

Villamil, O. S., & de Guerrero, M. C. M. (1996). Peer revision in the L2 classroom: Social-cognitive activities, mediating strategies, and aspects of social behavior. *Journal of Second Language Writing, 5*(1), 51–75. https://doi.org/10.1016/S1060-3743(96)90015-6

Wen, W., Charles, L., & Haggard, P. (2023). Metacognition and sense of agency. *Cognition*, *241*, 105622. https://doi.org/10.1016/j.cognition.2023.105622

Wiener, N. (1961). *Cybernetics: Or control and communication in the animal and the machine* (2nd rev. ed.). The MIT Press.

Williamson, B., Eynon, R., & Potter, J. (2020). Pandemic politics, pedagogies and practices: Digital technologies and distance education during the coronavirus emergency. *Learning, Media and Technology*, *45*(2), 107–114. https://doi.org/10.1080/17439884.2020.1761641

INDEX

New Literacies, New Learning: Cyber-Social Pedagogies in the Brazilian Context presents a journey of experimentation and discoveries about education in the digital environment, from the author's role as a History teacher to the completion of his doctoral research at the University of São Paolo (USP). The work explores how the digital ecosystem can promote critical, diverse, and inclusive education through a cyber-social design focused on Multiliteracies and new pedagogical practices.

Throughout six chapters, the book analyzes the transition from conventional educational environments to a cyber-social dynamic, where humans can harness the technological structure to create new learning opportunities. The author presents case studies conducted with the CGScholar platform, developed by the University of Illinois Urbana-Champaign and implemented in Brazil in partnership with the Faculty of Philosophy, Languages and Human Sciences (FFLCH-USP).

The work highlights innovative pedagogical resources, such as peer learning, process-oriented assessment, and recursive feedback, demonstrating how these practices can transform the teacher-student relationship into a more horizontal and collaborative experience. It also addresses the role of Artificial Intelligence in education and discusses the challenges of implementing these cyber-social ecosystems in the Brazilian context.

Despite Brazil's social and economic inequalities, the book argues that there are sufficient digital, theoretical, and empirical resources to invest in this path, reimagining our relationship with technology and engaging digital natives in meaningful learning that impacts their real-world experiences.

Rodrigo Abrantes da Silva holds a Ph.D. in Literacies from the University of São Paulo (USP). Guided by a multiliteracies framework, his research focuses on designing digital learning environments that embed pedagogical principles to empower students across diverse educational contexts. His professional experience spans secondary and higher education as well as teacher education programs. His work has received international recognition, including two International Excellence Awards for the papers *"Decolonial Practices on the CGScholar Educational Platform: Subjectification, Ecology of Knowledge, and the Design of Rhizomatic Multimodal Texts"* and *"New Digital Multiliteracies as a Learning Model: Fostering Collaboration, Identity, and Recognition."*

ISBN 978-1-966214-62-

9 781966 214625

e-Learning & Innovative
Pedagogies